Plant Sciences and Genetics in Agriculture

Plant Sciences and Genetics in Agriculture

Nirmay Adiga

RANDOM PUBLICATIONS
NEW DELHI (INDIA)

Plant Sciences and Genetics in Agriculture

ISBN 978-93-5111-372-0

Published in 2014 in India by

RANDOM PUBLICATIONS

4376-A/4B, Gali Murari Lal, Ansari Road
New Delhi-110 002
Phone : +91-11-43580356, +91-11-23289044
e-mail: randomexports@gmail.com, sales@randompublications.com, info@randompublications.com

Type Setting by: Friends Media, Delhi-110089
Printed at : Mehra Printers, Delhi-110092

Preface

Plant genetics is a very broad term. There are many facets of genetics in general, and of course there are many facets to plants. The definition of genetics is the branch of biology that deals with heredity, especially the mechanisms of hereditary transmission and the variation of inherited characteristics among similar or related organisms. And the definition of a plant is any of various photosynthetic, eukaryotic, multicellular organisms of the kingdom Plantae characteristically producing embryos, containing chloroplasts, having cell walls which contain cellulose, and lacking the power of locomotion. Although there has been a revolution in the biological sciences in the past twenty years, there is still a great deal that remains to be discovered. The completion of the sequencing of the genomes of rice and some agriculturally and scientifically important plants (for example Physcomitrella patens) has increased the possibilities of plant genetic research immeasurably. Plant genetics is different from that of animals in a few ways. Like mitochondria, chloroplasts have their own DNA, complicating pedigrees somewhat. Like animals, plants have somatic mutations regularly, but these mutations can contribute to the germ line with ease, since flowers develop at the ends of branches composed of somatic cells. People have known of this for centuries, and mutant branches are called "sports". If the fruit on the sport is economically desirable, a new cultivar may be obtained.

Some plant species are capable of self-fertilization, and some are nearly exclusively self-fertilizers. This means that a plant can be both mother and father to its offspring, a rare occurrence in the animals. Scientists and hobbyists attempting to make crosses between different plants must take special measures to prevent the plants from self-fertilizing. Plants are generally more capable of surviving, and indeed flourishing, as polyploids. Polyploidy, the presence of extra sets of chromosomes, is not usually compatible with life in animals. In plants, polyploid individuals are created frequently by a variety of processes, and once created usually cannot cross back to the parental type. Polyploid individuals, if capable of self-fertilizing, can give rise to a new genetically distinct lineage, which can be the start of a new species. This is often called "instant speciation". Polyploids generally have larger fruit, an economically desirable trait, and many human food crops, including wheat, maize, potatoes, peanuts, trawberries and tobacco, are either accidentally or deliberately created polyploids. Hybrids between plant species are easy to create by hand-pollination, and may be more successful on average than hybrids between animal species. Often tens of thousands of offspring from a single cross are raised and tested to obtain a single individual with desired characteristics. People create hybrids for economic and aesthetic reasons, especially with orchids. According to Vaughan A. Hilder and Donald Boulter at the Department of Biological Sciences, University of Durham, there have been serious failures in resistance to targeted pests in Bt cotton; most plant-derived resistance factors produce chronic rather than acute effects; and many serious pests are simply not susceptible to known resistance factors. According to John E. Berringer the outcome of releasing genetically modified organisms into the environment is still not known.

For every student, teacher and researcher in the subject it offers a solid basis for an in-depth understanding of the entire subject area.

I thank all members of my team who have helped in the preparation of the book. My special thanks go to "Random Publication" who have published the book.

—*Nirmay Adiga*

Contents

1

Applications of Cell Culture in Plant

Induced mutagens produce more frequency of mutants than spontaneous ones and screening them at cellular level also inhibits chimeric formation which is a drawback in mutation breeding. Plants being important source of variety of chemicals used in pharmacy, medicine and industry, cell cultures are effectively utilized for production of these chemicals on a commercial scale for enhanced yield and better production control.

Micropropagation

In vitro propagation of plants vegetatively by tissue culture to produce genetically similar copies of a cultivar is referred to as micropropagation or clonal propagation. Sexually propagated plants (through generation of seeds) demonstrate a high amount of heterogeneity since their seed progenies are not true to type whereas asexual reproduction (by multiplication of vegetative parts) gives rise to genetically identical copies of parent plant. Thus, it permits perpetuation of the parental characters of the cultivars among the plants resulting from micropropagation.

Micropropagation proves useful for propagation of:

- Sexually sterile species like triploids, aneuploids which cannot be perpetuated by seeds.
- Seedless plants like banana
- Cross bred perennials where heterozygosity is to be maintained
- Mutant lines like auxotrophs which cannot be propagated *in vivo*
- Disease free planting material of fruit trees and ornamentals.

Micropropagation involves following major stages:

- Stage 0 *Selection and maintenance of stock plants for culture initiation* (3 months).
- Stage I *Preparation and establishment of explant on suitable culture medium* (3-24 months) (usually shoot tips and axillary buds used).
- Stage II *Regeneration: multiplication of shoots or somatic embryos on defined* (10-36 months) *culture medium.*
- Stage III *Rooting of regenerated shoots/ somatic embryo in vitro* (1-6 weeks).
- Stage IV *Transfer of plantlets to sterilized soil for hardening under greenhouse environment.*
- (Stage III can be skipped for in vivo rooting of stage II regenerated shoots).

Advantages of micropropagation over conventional propagation methods ;

- Genotype constitution maintained as there is lesser variation in somatic embryo
- Easier transport and storage is facilitated by small size propagules and their ability to grow in soil less medium.
- Control over growing conditions as the production of planting material is completely under artificial control *in vitro*
- Reduced growth cycle and rapid multiplication as shoot multiplication has short cycle and each cycle results in exponential increase in number of shoots
- Selective multiplication can be done for e.g. auxotrophs, aneuploids, selected sex in dioecious species
- Virus free plants can be raised and maintained through meristem culture which is the only method available for this.

Disadvantages:

- Involves high cost
- Somaclonal variation- any variation if occurs during multiplication may go unnoticed
- Recalcitrancy of species/ genotype- many tree sp like mango etc. do not respond to *in vitro* growth.

Application of Micropropagation

1. Commercial production of secondary metabolites: The compounds/ biochemicals which are not directly involved in primary metabolic processes like respiration, photosynthesis etc. are secondary metabolites. These include a variety of compounds like alkaloids, terpenoids, phenyl propanoids etc. with various biological activities like antimicrobial, antibiotic, insecticidal, valuable pharmacological and pharmaceutical activities. Therefore, micropropagation allows their commercial scale production from cell cultures *viz.* shikonin derivatives used in dyes, pharmaceuticals are produced from cell cultures of *Lithospermum erythrorhizon*. Also, cultured cells of many plant species produce novel biochemicals which have otherwise not been detected in whole plants.
2. Production of synthetic seeds: Synthetic seed is a bead of gel containing somatic embryo or shoot bud with growth regulator, nutrients, fungicides, pesticides etc. needed for development of complete plantlet. These are better propagules as donot need hardening and can be sown directly in field.
3. Raising somaclonal variants: The genetic variability occurring in somatic cells, plants produced in vitro by tissue culture are referred to as somaclonal. When these variations involve traits of economic importance, these are raised and maintained by micropropagation.
4. Production of disease free plants: Most of the horticultural fruit and ornamental crops are infected by fungal, viral, bacterial diseases. Micropropagation provides a rapid method for production of pathogen free plants. In case of viral diseases especially, the apical meristems of infected plants are free or carry very low concentrations of viruses. Thus culturing meristem tips provides disease free plants.

Micro-propagation Methods: The ability of mature cell to dedifferentiate into callus tissue and the technique of cloning isolated single cell *in vitro* discussed earlier in this chapter have demonstrated that the somatic cells can differentiate to a whole plant under particular conditions. This potential of cell to divide and develop into multicellular plant is termed as cellular totipotency. To express totipotency, after dedifferentiation, the cell has to undergo redifferentiation or regeneration which is the ability of dedifferentiated cell to form plant or plant organs. This may occur through either of two processes:

- Organogenesis
- Embryogenesis.

Organogenesis is a process involving redifferentiation of meristematic cells present in callus into shoot buds. These shoot buds are monopolar structures which in turn give rise to leaf primordial and the apical meristem. The buds have procambial strands connected with preexisting vascular tissue present in the explant or callus. The stimulation of shoot bud differentiation in plants depends on many factors which differ for different plant species. In general, it is promoted by cytokinin and inhibited by auxins. The classical studies of Skoog and Miller (1957) demonstrated that the relative ratio of CK and auxin is important in determining nature of organogenesis in tobacco pith tissue. In tobacco, high level of CK initiates bud formation while high concentration of auxin favours rooting.

But there have been studies in other plant species which donot follow this concept of auxin/CK ratio. In most cereals, callus tissue exhibits organogenesis when it is subcultured from a medium containing 2,4-D to a medium where 2,4-D is replaced by IAA or NAA. GA3, which in general has inhibitory effect on shoot buds whereas many species show enhanced shoot regeneration due to abscissic acid. The variable responses of different plant species to the growth regulators is because the requirement of exogenous GRs depends on their endogenous levels which might differ in different plant species and also in different plant materials.

Other factors affecting organnogenesis are size and source of the explant. The larger the explant (containing parenchyma, cambium and vascular tissue), more is likelihood of shoot bud formation. Also, genotype of explant affects shoot regeneration as explant taken from different plant varieties of same species show different frequencies of shoot bud differentiation. Light has been shown to have inhibitory effect. Even the quality of light has effect as blue light has been shown to induce shoot formation and root by red light in tobacco. The optimum temperature required may vary with plant species.

Somatic embryogenesis: is a process involving redifferentiation of meristematic cells into nonzygotic somatic embryo which are capable of germinating to form complete plants. Somatic embryos are bipolar structures with radical and plumule is contrast to monopolar shoot bud with only plumular end in organogenesis. While developing into somatic embryo, the mersitematic cells break any cytoplasmic or vascular connections with other cells around it and become isolated. Therefore, unlike shoot bud, the somatic embryos are easily separable from explants. Somatic embryogenesis involves three distinct steps which are absent in organogenesis:

- Induction: is the initiative phase where cells of callus are induced to divide and differentiate into groups of meristematic cells called embryogenic clumps (ECs). These ECs develop into initial stages of somatic embryo i.e. globular stage.
- Maturation: In this phase somatic embryos develop into mature embryos by differentiating from globular to heart shaped, torpedo to cotyledonary stages. The mature embryo here undergoes biochemical changes to acquire hardiness.
- Conversion: Embryos germinate to produce seedlings.

Somatic embryogenesis is influenced by following factors:

Growth regulators: The presence of auxin (generally 2,4-D) in the medium is essential for induction phase. 2,4-D induces dedifferentiation of explant cells to form ECs. When auxin is removed or its concentration is reduced, ECs convert to somatic embryos. Once induced, cells don't need PGRs. Still some doses of CK at maturation and conversion make better plants. Maturation is achieved by culturing somatic embryos on high sucrose medium. Also, ABA is added as it gives hardening due to water loss which is important for embryo maturation. Ethylene inhibits both somatic embryogenesis and organogenesis. Therefore, silver nitrate is added to the medium as inhibitor of ethylene for plant regeneration.

- Nitrogen source: NH4+ form of nitrogen is essential for induction of somatic embryogenesis while NO3- form is required during maturation phase.
- Other factors: Like shoot bud differentiation, explant genotype has influence on somatic embryogenesis also. In cereals, use of maltose as carbohydrate source promotes both somatic embryo induction and maturation.

Anther Culture

Culturing anther on a suitable media to regenerate into haploid plants is called anther culture. First time, haploid plants were discovered in *Datura stramonium* by A.D. Bergner in 1921. Guha and Maheshwari (1964) pioneered the formation of embryos from anthers of *Datura innoxia* grown *in vitro*. After this, haploid plants have been produced via anther culture in more than 170 species. The anther culture technique is useful in haploid production. Haploid production: Haploid plant is defined as a sporophyte with gametophytic chromosome number. The *in vitro* production of haploid plants can be achieved by many techniques like:

- Delayed pollination which may not result in fertilization and hence only female genome grows up to form a haploid plant.
- Temperature shock – Extremes of temperature (both high and low) are used to suppress syngamy or make pollen inactive, thus leading to induction of haploidy.
- Irradiation effect - X rays, UV rays induce chromosomal breakage in pollen cells thus making them sterile which in turn results in haploid production.
- Chemical treatment – Treatment with colchicines, maleic hydrazide and toluene blue etc. also induces chromosomal elimination.
- Genome elimination by distant hybridization – In case of distant crosses like inter-generic and inter-specific crosses where during the developmental process, one of the parental genomes is selectively eliminated subsequently leading to formation of haploid plants.

Therefore, production of a haploid plant where egg cell is inactivated and only male genome is present is called androgenesis. Similarly, production of haploid by development of unfertilized egg cell due to inactivation of pollen is called gynogenesis. Among all the methods illustrated above, anther culture is the most popular and successful for haploid production.

Anther Culture Procedure

Step1 : Experimental material: Young healthy plants grown under controlled conditions are used as experimental material from which flower bud of right stage (varies with species) is excised.

Step2 : Disinfestation, excision and culture of anther: Flower buds are surface sterilized in laminar flow chamber followed by excision of anther from the bud. Stage of pollen development is determined by squashing an anther in acetocarmine and observing it under microscope. While excising anthers from flower buds, care is taken that anthers are not injured as injury leads to callusing hence giving mixture of diploids, haploids and aneuploids.

Step 3 : Culture medium conditions: The anthers are generally cultured on a solid agar medium where they develop into embryoids for anther culture under alternate light and dark period. Medium should have sucrose for induction of embryogenesis.

Step 4: Haploid plants: In species following direct androgenesis i.e. which develop through embryoid formation, small plants emerge in 3-8 weeks after culturing which are then transferred on to a rooting medium with low salt and small amount of auxin. Those species undergoing indirect androgenesis involving callus formation, callus is removed from the anther and placed onto a regeneration medium with suitable ratio of cytokinin to auxin. The haploid plants thus produced in both cases are transplanted to soil in small pots and maintained under controlled conditions in greenhouse.

Diploidisation of haploid plants: Haploid plants produced from anther culture maintained *in vitro* can grow till the flowering stage but cannot be perpetuated. Since these plants are haploid and have only one set of homologous chromosomes of the diploid species, they cannot form viable gametes and hence no seed setting takes place for further perpetuation.

Therefore, it is necessary to double the chromosome number of haploids to obtain homozygous diploids or dihaploid plants followed by their transfer to culture medium for further growth.

Application of haploid production: Diploidisation of haploid plants result in rapid achievement of homozygous traits in doubled haplouds, hence these anther derived haploid plants have been used in breeding and improvement of crop species.

Production of homozygous lines: The most important use of haploids is the production of homozygous lines which may be used directly as cultivars or may be used in breeding programme. For e.g. doubled haploids have been used for rapid development of inbred lines in hybrid maize programme. The anthers from F1 hybrids of selected or desirable cross are excellent breeding material for raising anther derived homozygous plants or doubled haploids in which complementary parental characteristics are combined in one generation.

The doubled haploid plants are subjected to selection for superior plants. This approach is described as hybrid sorting where recombinant superior gametes are virtually being selected since the heterozygous gene combination in the F1 hybrid is transformed into homozygous combinations. Hybrid sorting reduces the time required for haploid breeding by 4-5 years as in conventional breeding by pedigree/ bulk method, the same requires ten years. Also, selection among DH lines reduces the size of breeding population.

Gametoclonal variation: The variation observed among haploid plants having gametic chromosome number developing from anther culture is called gametoclonal variation. Such variations resulting in desirable traits are subjected to selection at haploid level followed by diploidisation to get homozygous plants which can be released as new varieties.

Selection of desirable mutants: Haploids offer a system where even recessive mutations are expressed unlike diploids where they express only in segregrating single plant progeny in M2 generation.

Therefore, in several crops desirable mutants including traits like resistance to diseases, antibiotics, salts etc. have been isolated from haploids derived from anther culture. For e.g. tobacco mutants resistant to black shank disease and wheat lines resistant to scab (*Fusarium graminearum*) have been selected and used as improved cultivars.

Problems associated with haploid plants:

- Many species are not yet amenable for haploid production
- Deleterious mutations may be induced during *in vitro* phase.
- Plants having more or lesser than gametic chromosome number is also obtained which necessitates cytological analysis first.
- Occurrence of gametoclonal variation limits the use of anther derived embryos for genetic transformation.

The major advantages of cell culture systems over the conventional cultivation of whole plants are:

- Higher and quicker yields of product from very small amount of plant material needed to initiate the culture in contrast to large amounts of mature plant tissues processed to achieve low yields of final product, for e.g. the dry weight of shikonin produced from cell culture is 20% more than from plants.
- In case of plant material facing threat of extinction or are limited in supply like *L erythrorhizon, in vitro* production of secondary metabolites is saving option.
- Controlled environmental conditions in cell culture ensure continuous supply of metabolites. In conventional system, source plant may be seasonal, location specific and also subject to

environmental degradation. Also, *in vitro* culture of cells is more economical for those plants which take long to achieve maturity.

- Bioconversion: Low cost precursors are supplied as substrates to cell cultures for conversion to the high cost final product, thus minimizing labour, cost and time. Also, specific substrates can be biotransformed to more valuable product by single step reaction *in vitro.*
- Production of novel compounds: Mutants cell lines can be utilized to produce novel compounds which were not previously found naturally in plants *viz.*cell suspension cultures of *Rauwolfia serpentina* produce novel glucosides of ajmaline (alkaloid).

Production of Secondary Metabolites

Plants are the source of a large variety of biochemicals which are produced as both primary and secondary metabolites. Primary metabolites include nucleic acids, proteins, carbohydrates and fats which along with their intermediates function for survival of cell and organism.. Compounds like alkaloids, nonprotein aminoacids, terpenoids and phenolics are grouped under secondary metabolites which donot participate in vital metabolic function of cell. Primary metabolites essentially provide the basis for growth and reproduction, while secondary metabolites for adaptation and interaction with the environment. As secondary metabolites provide industrially important natural products like colour, insecticides, antimicrobials and fragrances, therapeutics etc., they are of great economic importance.

Therefore, plant tissue culture is being potentially used as an alternative to plants for production of secondary metabolites. The first large scale production was successfully done for shikonin produced from *Lithospermum erythrorhizon.*

It is used as antiseptic and as dye for cosmetics. Since then many valuable secondary metabolites like taxol, berberine etc. have been obtained using tissue culture. Increasing productivity of secondary metabolites by cell cultures: In plants, most of the secondary metabolites are produced in differentiated cells or organized tissues. However, callus and cell suspension culture lack organ differentiation and hence produce low yields of these biochemicals. The yield of secondary metabolite by undifferentiated tissue or cell cultures can be increased by following techniques:-

- Select proper cell line: The heterogeneity within the cell population can be screened to select lines capable of accumulating higher level of metabolite.
- Medium manipulation: The constituents of culture medium like nutrients, phytohormones also the culture condition like temperature, light etc. influence the production of metabolites.

  For e.g. if sucrose concentration is increased from 3% to 5%, the production of rosamarinic acid is increased by five times. In case of shikonin production, IAA enhances the yield whereas 2,4-D and NAA are inhibitory.
- Elicitors: Compounds that induce the production and accumulation of secondary metabolite in plants are known as elicitors. Elicitors produced within the plant cells include cell wall derived polysaccharides like pectin, pectic acid, cellulose etc.

  Product accumulation also occurs under stress caused by physical or chemical agents like UV, low or high temperature, antibiotics, salts of heavy metals, high salt concentration grouped under abiotic elicitors. These elicitors when added to medium in low concentration (50-250ng/l) enhance metabolite production.
- Permeabilisation: Secondary metabolites produced in cell are blocked in the vacuole. By manipulating the permeability of cell membrane, they can be elicited out to media. Permeabilisation can be achieved by electric pulse, UV, pressure, sonication, heat. Even charcoal is added to medium to absorb secondary metabolites.
- Immobilisation: Cell cultures encapsulated in agarose and Calcium alginate gels or entrapped in membranes are called immobilised plant cell culture. Here cell to cell contact is better while cells are also protected from high shear stresses.

  These immobilized systems effectively increase the productivity of secondary metabolites in number of species. Elicitors can also be added to these systems to stimulate secondary metabolism.

Limitations

- High production cost is involved.
- Lack of knowledge of biosynthetic pathways of many compounds is major bottleneck in improvement of their production.

- Cultured plant cells are often unable to produce high value compounds.

Protoplast Isolation and Fusion

A fundamental difference between plant and animal cells is that 'plant cells are totipotent' which formed the basis of plant tissue culture. Another important difference is that 'presence of cell wall in plants' paved way for the most significant development in the field of plant tissue culture which is isolation, culture and fusion of protoplasts. Plant cells from which cell wall has been removed are termed protoplasts. Cultured protoplasts besides being used for genetic recombination through somatic cell fusion can also be used for taking up foreign DNA, cell organelles, bacteria and virus particles. Therefore, protoplast culture has achieved important status in plant biotechnology.

Isolation of Protoplast: Plant protoplasts were first isolated by Klercker in 1892 from onion bulb scales plasmolysed in hypertonic solution. This mechanical procedure gave low yield of protoplasts and could be utilized for only highly vacuolated and non meristematic cells. It was in 1960 when Cocking demonstrated the isolation of intact protoplasts by use of cell wall degrading enzyme, cellulase prepared from the fungus, *Myrothecium verrucaria.* By 1968, commercial preparations of purified cell wall degrading enzymes such as macerozyme, cellulase and hemicellulose became available that gave further progress to enzymatic isolation of protoplasts. Enzymatic method of protoplast isolation can be classified into two groups:

1. Sequential enzymatic: This involves two steps where first macerated plant tissues are incubated with pectinase to get single cells followed by cellulase treatment to get protoplast.
2. Mixed enzymatic: This involves simultaneous separation of cells and degradation of their walls to convert protoplast by immersing plant tissues in mixture of pectinases and cellulases.

Protoplasts have been isolated from all plant parts like shoot tips, cotyledons, flower petals, however, leaf mesophyll tissue is widely preferred for its high reproducible potential for regeneration. Generally, the youngest, fully expanded leaves from young plants or seedlings are used as source of protoplast. Preconditioning plants in darkness or cold (4-10°C) for 24-72h before protoplast isolation improves protoplast yield. The leaves are surface sterilized and cut into small pieces and floated on an osmotically adjusted solution at 20-25°C for 1-24h. During this step, plasmodesmatal complex is broken down, water moves out of cell which causes the cell contents to shrink and draw away from the cell wall. This allows cells to retain their integrity after the cell wall is removed otherwise they start fusing. This step is followed by incubation of leaf pieces with digestive enzymes in darkness on shaker (30-50rpm). Incubation time and temperature varies with species and time. The osmotic concentration of enzyme mixture and of subsequent media is elevated by adding sorbitol or mannitol to stabilize protoplasts or they will burst. Addition of 50-100mM/l CaCl2 improves stability of plasma membrane. At acidic pH in the range of 5.0-6.0 is optimal. The buffer often contains phosphate at 3mM to minimize shifts in pH during digestion. In case of monocots, where mesophyll cannot be the source, cell suspension cultures are used for isolation of protoplasts. The released protoplasts are separated by centrifugation at low speed for about ten minutes followed by washing two to three times before transfer to culture medium to remove enzymes and debris.

Protoplast viability and Plating Density: Before culturing protoplast, their viability is estimated by staining with Fluorescein diacetate (FDA). Viable protoplasts exhibit green fluorescence under UV. It should be examined within 5-15 min after the FDA treatment after FDA dissociates. Some protoplasts have leaky membrane through which metabolites are leached out. If sufficient density of protoplasts is there, metabolites can be absorbed back. Therefore, a minimum plating density is required for growth to begin and it is estimated by haemocytometer. For genetic engineering, single or fewer protoplast culture is required for which either conditional media or feeder layer are used. Conditional media is one where plant cells were already grown and so has metabolites leached into it. After filtering, this media is used for growing isolated protoplasts. Feeder layer is prepared by plating solid media with protoplasts followed by irradiation which inactivates the nucleus but protoplasts are viable. Protoplasts at lower densty can now be plated on this feeder layer.

Protoplast Culture: Freshly isolated protoplasts are spherical because they are unbound by cell wall. Viable protoplasts regenerate a new cell wall within 48 to 96 h after isolation which can be determined by staining with calcafluor white. Protoplasts with new cell wall fluoresce bluish white under UV. Protoplasts that fail to regenerate a wall generally will not divide and die eventually. Also, all the healthy protoplasts may not divide and therefore, plating efficiency is calculated to estimate cell vigor. Plating efficiency is number of dividing protoplasts/total number of protoplasts plated. The protoplasts capable of dividing undergo first division within 2-7 days after isolation. The delicate nature of protoplasts demand modifications in MS and B5 media or any other culture medium used for organ regeneration from explants. Besides higher osmotica, the inorganic salt concentration is adjusted (Ammonium nitrate concentration is lowered and calcium level is increased), more of organic components, vitamins and PGRs are added to hasten and promote cell wall synthesis. Due to sensitivity to light, protoplasts are cultured in diffuse light for initial 4-7 days. After early culture, when protoplasts have regenerated new cell wall and divided, they are transferred back to normal medium where plants regenerate by shoot formation or somatic embryogenesis. Plant regeneration from protoplasts has been reported from alfalfa, tobacco, carrot, tomato etc. but cereals still pose problems.

Protoplast Fusion and Somatic Hybridization: Purified protoplasts once obtained from any two different sources (can be different tissues, different plants or species or different genera), they can be fused together to form somatic hybrids. This non-conventional method of genetic recombination involving protoplast fusion under *in vitro* conditions and subsequent development of their product to a hybrid plant is known as somatic hybridization. First, somatic hybrid plant of *Nicotiana glauca* (+) *N. langsdorfii* was reported by Carlson in 1972.

Protoplasts can be induced to fuse by variety of fusogens or electrical manipulations which induce membrane instability. Most commonly reported fusion inducing agents are sodium nitrate (used by Carlson), high pH/Ca2+ concentration and Polyethylene glycol (PEG) treatment. Sodium nitrate treatment results in low frequency of heterokaryon formation, high pH and high Ca2+ concentration suits few plant species whereas PEG is the most favoured fusogen for its reproducible high frequency of heterokaryon formation and low toxicity. However, treatment with PEG in presence of high pH/ Ca2+ is reported to be most effective in enhancing heterokaryon formation and their survivability.

A more selective, simpler, quick and non toxic approach is electrofusion which utilizes electric shock or short pulse of high voltage to promote membrane fusion between two cells. Many useful somatic hybrid plants produced by electrofusion have been reported like *Nicotiana plumbaginifolia* (+) *N tobacum*, *Solanum tuberosum* (+) *S. charcoense* (resistant to Colorado potato beetle).

Somatic Hybrids

Protoplast population following fusion treatment is heterogeneous mixture of unfused parents, homokaryons, heterokaryons, fused protoplasts with independent two nuclei etc. thus necessitating selection of stable hybrids. Therefore, some identification and selection system should be incorporated into each parental cell line before fusion. Usually, two parental cell lines with differing requirements as selective screens make selection of hybrid convenient as only those fused cells that possess complementary traits of both parents will thrive. Resistance to antibiotics, herbicides and ability to grow on specific amino acid analogs are few of the selection methods used. Besides these, two different vital stains which donot affect viability of cells like fluorescein isothiocynate (FITC) and rhodomine isothiocyanate (RITC) have been successfully used. Under fluorescent microscope, FITC stained cells appear green and RITC stained protoplasts appear red whereas fused cells having both RITC and FITC fluoresce yellow.

Application of Somatic Hybridization

- Genetic recombination in asexual or sterile plants: Protoplast fusion has overcome the impediment of reproduction in haploid, triploid and aneuploid plants. Also, genomes of asexually reproducing plants can be recombined using this approach *viz.* protoplasts isolated from dihaploid potato clones have been fused with protoplasts of *S. brevidens* to produce hybrids of practical breeding value.
- Genetic recombination between sexually incompatible species: The incompatibility barriers in sexual recombination at interspecific or intergeneric levels are also overcome by somatic hybrisation. Generally, somatic hybrids are used for transfer of useful genes such as disease resistance, abiotic stress resistance or genes of industrial use for e.g. Datura hybrids (*D. innoxia* + *D discolor*, *D. innoxia* + *D stramonium*) show heterosis for scopolamine (alkaloid) content which is 20-25% higher than in parent species and therefore has industrial application.
- Cytoplasm transfer: Somatic hybridization minimizes the time taken for cytoplasm transfer to one year from 6-7 years required in back cross method. Also, this method allows cytoplasm transfer between sexually incompatible species. Cybrids have cytoplasm from both parents but nucleus of only one. Nucleus

of other parent is irradiated. This approach has been potentially used to transfer two desirable traits – cytoplasmic male sterility (CMS) and resistance to atrazine herbicide, both coded by cytoplasmic genes in Brassica to different crops like tobacco, rice etc.

Somaclonal Variation

The genetic variability present in somatic cells, plants or plant progenies derived from cells/tissue cultured in vitro is called somaclonal variation. Larkin and Scrowcroft (1981) coined this term for all plant variants derived from any form of cell or tissue culture. Some variants are obtained in homozygous condition in the plants regenerated from cells cultured *in vitro* (R0 generation) but mostly variants are recovered in the selfed progeny of tissue culture regenerated plants (R1 generation).

This variation includes aneuploids, sterile plants and morphological variants, some of which may involve traits of economic importance for crop plants. Somaclonal variation may be genetic or epigenetic. Since only gametic variation follow Mendelian inheritance pattern and transmitted to next generation, they are important for crop improvement. Therefore, in several crops, R0, R1 and R2 progenies are analysed for transmission of variant trait to sexual progeny (R1) and 3:1 segregation leading to isolation of true breeding variants (R2).

The significance of somaclonal variation in crop improvement was first demonstrated in sugarcane and potato where few somaclones with disease resistance against Fiji, downy mildew (sugarcane) and late and early blight in case of potato were recovered.

Major Causes of Somaclonal Variation

Physiological: Variations induced by physiological factors in culture medium for e.g. prolonged exposure to PGRs (2,4-D; 2,4,5-T) results in variability among the regenerants. Often such variations are epigenetic and hence donot follow Mendelian inheritance.

Genetic: All the alterations at chromosomal level are grouped under genetic cause of variation observed in regenerants. Chromosomal rearrangements such as deletion, duplication, translocation, inversion polyploidy, aneuploidy, have been reported to be the chief source of somaclonal variation. Meiotic crossing over involving symmetric and asymmetric recombination could also be responsible for variation observed among somaclones.

Transposable elements like Ac-Ds in maize have been shown to get activated in *in vitro* culture. In maize (*Zea mays* L.) and broad beans (*Vicia faba* L), late replicating heterochromatin is the main cause of somaclonal variation. Single gene mutations in cultures also give rise to variations which are not detected in plants regenerated *in vitro* from any cell or tissue (R0 plants) but express in R1 plants (after selfing R0 plants).

Biochemical: The most common kind of biochemical variation is change in carbon metabolism leading to failure of photosynthesis *viz.* albinos is rice. Any variation in other cell processes like starch biosynthesis, carotenoid pathway, nitrogen metabolism, antibiotic resistance etc. also lead to somaclonal varaiation.

Application in Crop Improvement: Somaclonal variation represent useful source for introduction of valuable variations to plant breeders. Cell culture systems are well defined controlled environments, away from limitations of availability of space, time and variations due to environmental effects which are major bottlenecks in conventional mutation breeding. Somaclonal variation occurs at much higher frequencies than induced mutants which are associated with undesirable features. Cell culture systems allow plant breeder to have greater control on selection process as here they have the option to select from large amount of genetically uniform material.

Therefore, this is the only approach for genetic improvement in perennial species limited by narrow germplasm and long regeneration cycle, asexually propagated plants like bananas, for isolation of biochemical mutants like auxotrophs. Somaclonal variants have been isolated for variety of valuable traits like disease resistance, stress (salt, low temperature) resistance, improved yield and efficient nutrient uptake etc. Bio-13 is a somaclonal variant of *Citronella java,* a medicinal plant which yields 37% more oil and 39% more citronellol than the control varieties. Pusa Jai Kisan, with bolder seeds and higher yield developed at Indian Agricultural Research Institute is another successful application of somaclonal variation of *Brassica juncea* variety 'Varuna'. Somaclonal variants in Lathyrus with low toxin level have also been developed at Indian Agricultural Research Institute.

Germplasm Conservation

The genetic material especially its molecular and chemical constitution that is inherited and transmitted from one generation to other is referred to as germplasm. In other words, the sum total of all the genes present in a crop and its related species constitutes its germplasm. It is generally represented by a collection of various strains and species. Germplasm is valuable because it contains diversity of genotypes that is needed to develop new and improved genetic stocks, varieties and hybrids. Therefore, germplasm is the basic indispensable ingredient of all breeding programmes and great emphasis is placed on collection, evaluation and conservation of germplasm. The continuing search for high yielding varieties of crop plants with resistance to biotic and abiotic stresses necessitates the conservation of germplasm of different crops and their wild and weedy relatives.

In-situ conservation: *In situ* (on-site) conservation refers to the maintenance and use of wild plant populations in the habitats where they naturally occur and have evolved without the help of human beings. The wild populations regenerate naturally and are also dispersed naturally by wild animals, winds and in water courses. There exists an intricate relationship, often interdependence, between the different species and other components of the environment (such as their pests and diseases) in which they occur. The evolution is purely driven by environmental pressures and any changes in one component affect the other. Provided that changes are not too drastic, this dynamic co-evolution leads to greater diversity and better adapted germplasm. The conservation of the forests and other wild plant species is often carried out through protected areas such as national parks, gene sanctuary and nature reserves. However, this mode of conservation has certain limitations such as there is risk of loss of material due to environmental hazards.

Ex-situ conservation: Ex situ (off-site) conservation of germplasm takes place outside the natural habitat or outside the production system, in facilities specifically created for this purpose. This is the chief mode of preservation of genetic resources for both cultivated and wild material. The most convenient method of ex-situ germplasm conservation is in the form of seeds. Thus, majority of field crops and vegetables which produce orthodox (dessication tolerant) seeds are conserved in gene banks by reducing their moisture content (3-7%) and storing under low humidity and low temperature. In case of crops with dessication sensitive or recalcitrant seeds (which lose their viability after being dried below a critical limit) and also in vegetatively propagated crops, *in vitro* methods are the most useful for germplasm conservation. This tissue culture based method has been mainly utilized for conservation of somaclonal and gametoclonal variations in cultures, plant material from endangered sp., plants of medicinal value, storage of pollen, storage of meristem culture for production of disease free plants and genetically engineered materials.

In vitro Germplasm conservation: Germplasm can be stored in vitro in variety of forms including isolated protoplasts, cells from suspension or callus cultures, meristem tips, somatic embryos, shoot tips or propagules at various stages of development. Methods for *in vitro* germplasm conservation are classified into two groups based on culture growth:-

1. Slow growth cultures: where limited growth of culture is allowed. This is a simple, effective and economic method and can be used in all species where shoot tip/ nodal explant are available. In these techniques, growth is suspended by either cold storage or lowering oxygen concentration. Such methods require serial subculturing for periodic renewal of cultures. The storage of germplasm by repeated cultures has some disadvantages like during subculturing there is risk of contamination by pathogen, genetic changes may also occur.
2. Cryopreservation: Any growth in plant cell and tissue culture is brought to a halt still retaining its viability in this technique by storing at ultra low temperature (-196°C) using liquid nitrogen. This method, also called freezepreservation, is most popular and effective for indefinite storage. Cryopreservation for germplasm purposes utilizes shoot tips and buds only but protoplasts, cells, tissues and somatic embryos are also cryopreserved for other tissue culture processes.

Factors affecting viability of cells frozen for cryopreservation:

- Physiological state of material: Cells in the late lag or exponential phase are considered ideal for freeze preservation. After thawing, these cytoplasm rich cells are able to retain their viability and grow again from the actively dividing meristematic cell component. But in shoot tips, embryos etc., tissue is large with highly vacuolated cells which get damaged by freezing and are unable to recover back.
- Prefreezing treatment: Conditioning treatment given to cells before freezing results in their hardening and increased survival rates. Such hardening treatments include growing culture in presence of cryoprotectant or growing at low temperature (4°C)

(for cold dormant sp) or in presence of osmotic agents like sucrose. These treatments function by either changing the cell water content, metabolite content or membrane permeabilites.

- Cryoprotectants: are chemicals imparting protection to withstand low temperature. For plants, most frequently used cryoprotectant is Dimethyl sulphoxide (DMSO). About 5-10% of DMSO is prepared and added gradually to prevent plasmolysis of the cells. Other commonly used cryoprotectants include glycerol, polyvinyl pyrollidone, polyethylene glycol (PEG) etc.
- Thawing rate and reculture: For better survival of preserved samples, rapid thawing from -196°C to about 22°C is recommended. By thawing rapidly, the damaging effects of ice crystal formation (crystallization of cell water while freezing) are minimized. These thawed samples during reculturing require special growth conditions, for enhanced recovery rates like dim light, high osmoticum, gibbrellic acid, and activated charcoal in the medium.

Methods of Cryopreservation

The sensitivity of cells to low temperature varies with the species. However, usually the sample to be preserved are treated with suitable cryoprotectant and then frozen by any one of the following methods:

- Rapid freezing: The vials with plant materials are directly dipped in liquid nitrogen. The temperature lowers very fast at the rate of 200°C/minute. It is a very hard treatment and hence survival rate is low. However, this method has been successful for germplasm conservation of large number of species where plant material with small size and low water content has been chosen.
- Controlled freezing: The plant material is cooled stepwise from room temperature to intermediate temperature (-20°C) maintained at that temperature for thirty minutes followed with rapid freezing by dipping into liquid nitrogen. This is a reliable method and is applicable to wide range of plant materials including shoot apices, buds and suspension cultures.

Advantages of Cryopreservation

- Indefinite preservation as metabolism comes to halt
- Low maintenance as only liquid nitrogen needs to be replenished

- No contamination
- Applicable to all species amenable to tissue culture

Limitations:

- Sophisticated equipment and facilities required
- Expertise needed
- Cells/tissues get damaged due to ice crystal formation or high solute concentration during dessication.

Genetic Engineering and Gene Technology

Since the 1970s, there have been considerable developments in the methods and techniques used to study biological processes at the molecular level. The discovery of restriction enzymes, transformation technique within this period were breakthrough developments which allowed scientists to cut, splice and alter DNA, the molecule that carries the blueprints for life. This further led to isolation of single DNA sequence from one organism by breakage of DNA molecule at two desirable places and then inserts it at a desired position in another DNA molecule from completely different organism to form recombinant DNA and the technique involved is called recombinant DNA technique. This has also been termed as genetic engineering because of the potential for creating novel genetic combinations. Using this technique, single copy of a gene or DNA can be isolated and cloned into indefinite number of identical copies and this is known as gene cloning.

The basic steps involved in recombinant DNA technique can be outlined as follows:

1. Isolation of desired DNA fragment or gene of interest
2. Insertion of the isolated gene in a suitable vector.
3. Introduction of this recombinant molecule into host cell by transformation
4. Selection of transformed host cells
5. Multiplication and expression of introduced gene in the host.

To explain these steps of gene cloning, understanding of a few basic techniques is essential which are discussed below.

Gel Electrophoresis: Gel electrophoresis is a technique used to separate molecules based on physical characteristics such as size, shape, or isoelectric point. A gel is composed of agarose or polyacrylamide. Agarose is used for separating DNA fragments in size range of 100bp to 20kb. Polyacrylamide is preferred for smaller DNA fragments. Separation is achieved by moving the negatively charged nucleic acid molecules through agarose matrix with an electric field. Rate of migration of DNA molecules is inversely proportional to their molecular weight. Smaller molecules move faster than larger ones. But conformation of the DNA molecule is also an important factor. Conformations of a DNA plasmid that has not been cut with a restriction enzyme will move with different speeds such as: supercoiled plasmid DNA will move fastest followed by linearised DNA, while open circular DNA will move at the slowest pace. To visualise DNA, gels are stained with ethidium bromide which intercalates in DNA and fluorescse under UV.

Restriction enzymes: Restriction endonucleases are enzymes that cleave the phosphodiester bond within a polynucleotide chain. The chemical bonds that the enzymes cleave can be reformed by other enzymes known as ligases, so that restriction fragments generated from different genes can be joined together These enzymes cleave DNA at specific sites and their recognition sequences or have two fold axis of rotational symmetry or are palindrmic (i.e. the sequence on one strand reads the same in the same direction on the complementary strand).

Polymerase chain reaction: Polymerase chain reaction is an extremely versatile technique for copying DNA. In brief, PCR allows a single DNA sequence to be copied to millions of times in predetermined ways. It usually consists of a series of 20 to 35 cycles. Commonly, PCR is carried out in three steps such as denaturation, primer annealing followed by primer extension.

1. Prior to the first cycle, during an initialization step, the PCR reaction is often heated to a temperature of 94-96°C to ensure that most of the DNA template and primers are denatured. Also, some PCR polymerases require this step for activation. Following this hold, cycling begins, with denaturation step at 94-96°C for 20-30 seconds.
2. The denaturation is followed by the annealing step. In this step the reaction temperature is lowered so that the primers can attach to the single-stranded DNA template. The temperature at this step depends on the Tm of the primers and is usually between 50-64°C for 20-40 seconds.
3. The annealing step is followed by an extension/elongation step during which the DNA polymerase copies the DNA template, starting at the primers annealed to both of its strands. The

temperature at this step depends on the DNA polymerase used. Taq polymerase has a temperature optimum of 70-74°C; thus, in most cases, during the extension a temperature of 72°C is used.

PCR can be used for lot of molecular biology experiments such as to introduce restriction enzyme sites, or to mutate (change) particular bases of DNA. PCR can also be used to determine whether a particular DNA fragment is found in a cDNA library. PCR has many variations, like reverse transcription PCR (RT-PCR) for amplification of RNA, and, more recently, real-time PCR (qPCR) which allow for quantitative measurement of DNA or RNA molecules.

Southern bloting: It is a method routinely used in molecular biology to check for the presence of a DNA sequence in a DNA sample. Southern blotting combines agarose gel electrophoresis for size separation of DNA with methods to transfer the size-separated DNA to a filter membrane for probe hybridization. The method is named after its inventor, the Edwin Southern. In this method, DNA samples are separated by gel electrophoresis and then transferred to a membrane by blotting via capilliary action. The membrane can then be probed using a labelled DNA complementary to the sequence of interest. Most original protocols used radioactive labels, however now non-radioactive alternatives are available. Southern blotting is less commonly used in laboratory science due to the capacity of using PCR to detect specific DNA sequences from DNA samples.

However, these blots are still used for some applications, such as measuring transgene copy number in transgenic plants. Similarly, northern blotting is a technique used in research to study gene expression. It takes its name from the similarity of the procedure to the Southern blot procedure but the key difference is, instead of DNA, RNA is used for blotting and analysis by electrophoresis and detection with a hybridization probe. A western blotting is a method to detect protein in a given sample of tissue homogenate or extract. The proteins transferred to membranes are "probed" using antibodies specific to the protein.

DNA Sequencing

The term DNA Sequencing encompasses biochemical methods for determining the order of the nucleotide bases, adenine, guanine, cytosine, and thymine, in a DNA oligonucleotide. Currently there are two types of DNA sequencing methods:

1. Maxam-Gilbert sequencing:In 1976-1977, Allan Maxam and Walter Gilbert developed a DNA sequencing method based on chemical modification of DNA and subsequent cleavage at specific bases. The method requires radioactive labelling at one end and purification of the DNA fragment to be sequenced. Chemical treatment generates breaks at a small proportion of one or two of the four nucleotide bases in each of four reactions (G, A+G, C, C+T). Thus a series of labelled fragments is generated, from the radiolabelled end to the first 'cut' site in each molecule. The fragments are then size-separated by gel electrophoresis, with the four reactions arranged side by side. To visualize the fragments generated in each reaction, the gel is exposed to X-ray film for autoradiography, yielding an image of a series of dark 'bands' corresponding to the radiolabelled DNA fragments, from which the sequence is inferred.
2. Sanger and Coulson method: The chain-terminator method developed by Sanger was more efficient and rapidly became the method of choice. The Maxam-Gilbert technique requires the use of highly toxic chemicals, and large amounts of radiolabeled DNA, whereas the chain-terminator method uses fewer toxic chemicals and lower amounts of radioactivity. The key principle of the Sanger method was the use of dideoxynucleotides triphosphates (ddNTPs) as DNA chain terminators. The classical chain-termination method requires a single-stranded DNA template, a DNA primer, a DNA polymerase, radioactively or fluorescently labeled nucleotides, and modified nucleotides that terminate DNA strand elongation. The DNA sample is divided into four separate sequencing reactions, containing the four standard deoxynucleotides (dATP, dGTP, dCTP and dTTP) and the DNA polymerase. To each reaction is added only one of the four dideoxynucleotides (ddATP, ddGTP, ddCTP, or ddTTP). These dideoxynucleotides are the chain-terminating nucleotides, lacking a 3'-OH group required for the formation of a phosphodiester bond between two nucleotides during DNA strand elongation. Incorporation of a dideoxynucleotide into the nascent (elongating) DNA strand therefore terminates DNA strand extension, resulting in various DNA fragments of varying length. The dideoxynucleotides are added at lower concentration than the standard deoxynucleotides to allow strand elongation sufficient for sequence analysis.

The newly synthesized and labeled DNA fragments are heat denatured, and separated by size (with a resolution of just one nucleotide), by gel electrophoresis on a denaturing polyacrylamide-urea gel. Each of the four DNA synthesis reactions is run in one of four individual lanes (lanes A, T, G, C); the DNA bands are then visualized by autoradiography or UV light, and the DNA sequence can be directly read off the X-ray film or gel image. The terminal nucleotide base can be identified according to which dideoxynucleotide was added in the reaction giving that band. The relative positions of the different bands among the four lanes are then used to read the DNA sequence.

Gene Cloning

Isolation of the desired gene: The DNA fragment of interest can be obtained from any of the following:-

a. Genomic library: is a large collection of recombinants in plasmid or phage vector, so that sum total of DNA inserts in this collection represent the entire genome of the concerned organism. The genomic library is prepared by shotgun approach where the total genomic DNA of organism is extracted and subjected to partial digestion by any restriction endonuclease or by sonication. These partial digests are then separated on agarose gel by electrophoresis or by sucrose density gradient centrifugation for selection of appropriate size fragments. The selected fragments are then inserted into phage ë vector or cosmid vector since these vectors can take up DNA inserts of upto 23-25kilobase (kb) pairs. The recombinant molecules are then cloned in a suitable bacterial host thus constructing a genomic library. The clone having the desired DNA insert is identified by screening the library using suitable probe. The probe may be mRNA of the gene, complementary DNA (cDNA) of its mRNA, homologous gene from another organism or synthetic oligonucleotide representing the sequence of a part of the desired gene. Once identified, the desired clones are picked up from the library.

b. cDNA library: A cDNA library is a population of bacterial transformants or phage lysates in which mRNA isolated from tissue or organism is represented as its cDNA inserts. cDNA is complementary DNA produced by reverse transcriptase using mRNA as template. The isolation and identification of desired clone from cDNA library is done by screening in a similar manner to that of genomic library. However, cloned eukaryotic cDNAs have their own special uses since they lack intron sequences that are usually present in the corresponding genomic DNA. Introns are noncoding sequences that often occur within eukaryotic coding sequences (exons) and are sliced out while post transcriptional processing to produce mRNA. Since, bacteria donot possess the enzymes necessary for splicing of intron, eukaryotic cDNA clones become essential when the expression of eukaryotic gene is required in a prokaryote. Also, if the sequence of genomic DNA is known, the intron/exon boundaries can be assigned by comparison with the cDNA sequence.

c. Chemical synthesis of gene: The basic sequence of any gene can be deduced from the nucleotide sequence of mRNA or amino acid sequence of the protein coded by it. The polynucleotide of deduced base sequence can be synthesized chemically using automated DNA synthesizers.

d. Gene amplification through PCR: The polymerase chain reaction technique amplifies single copy of the desired DNA to billion copies in few hours. This PCR based approach is quicker and simpler than library construction and screening and hence preferred to all the aforesaid approaches.

Insertion of the isolated gene in a suitable vector: Vector is a DNA molecule into which exogenous DNA is integrated for cloning and that has the ability to replicate in a suitable host cell. Vectors are used to assist in the transfer, replication and sometimes expression of a specific DNA sequences in a target cell. Therefore, vector must have the following properties:

- Vector must have origin of replication to replicate autonomously in the cell population as the host organism grows and divides. Their maintenance should not necessarily require integration into host genome.
- Vector must have unique sites for many restriction enzymes called multicloning site (MCS) into which DNA insert can be cloned without disrupting as essential function.
- Vector must be fairly small, low molecular weight DNA molecules to facilitate their isolation and handling.
- Vector must have some selectable marker that will enable the recombinant vector to be selected from large population of cells that have not taken up foreign DNA.

A variety of vectors have been developed to function as cloning vehicles:

Plasmids are self replicating circular duplex DNA molecules which are stably inherited in an extrachromosomal state. Plasmid vectors are used for cloning DNA of small size (upto 12kb). The circular plasmid DNA which is to be used as vector is first cleaved by restriction endonuclease (RE) to give linear DNA molecule.

The foreign DNA to be inserted is also cut by same endonuclease foolowed by ligation (joining) of the linearised vector and insert DNA resulting in a bigger circular DNA that can now be separated by gel electrophoresis on the basis of its size.

Selection of the chimeric DNA can also be done if insertion of foreign DNA at a endonuclease site inactivates a gene whose phenotype is readily scorable.

Usually, the selectable marker is resistance to different antibiotics. The naturally occurring plasmids of *Escherichia coli* have been modified, shortened, reconstructed and recombined to create many different plasmids of enhanced utility as vectors for e.g. pBR322, pUC series (pUC18, pUC19, pUC8 etc).

Bacteriophage vector: are viruses of bacteria that either infect the cell and lyse them (lytic cycle) or integrate into bacterial chromosome and multiply with it as prophage (lysogenic cycle). They act as cloning vehicles for larger pieces of DNA (23-25kb) as compared to plasmid vectors. Also, screening of recombinants is easier as phages form plaques (a clear zone where lysis has occurred in the bacterial lawn) in contrast to plasmids where bacterial colonies are screened. Most commonly used phage vector are lambda (λ) and M13 phages.

Cosmid vectors: are plasmids which contain a fragment of λ DNA including the *cos* site. Since cos site enables the DNA to get packed in λ particle *in vitro. Recombinant* cosmid DNA is injected and circularized like phage DNA which infects host cells which is more efficient than plasmid transformation. But it replicates as a plasmid without expression of any phage functions. With a cosmid vector of 5kb, large DNA inserts (32-47kb) are selected as the distance between two cos sites must be between 38 and 52kb for packaging.

Therefore, cosmids can accommodate upto 45kb long DNA inserts which is much more than a phage vector. Because of their capacity for taking up large fragments of DNA, cosmids are particularly used as vectors for constructing libraries of eukaryotic genome fragments.

Phagemid vector: Those vectors that have origin of replication derived from both a plasmid and phage are known as phagemids. Under normal circumstances, the plasmid *ori* is used for replication but following a phage infection the other ori is used and single stranded DNA is produced which is useful for sequencing. One such plasmid is pBluescript which has its MCS flanked by T3 and T7 promoters, enabling expression of the cloned insert to be obtained regardless of its orientation.

Phasmid vector: are plasmid vectors containing λ attachment site (λatt) and functional origins of replication of both plasmid and λ. λatt allows plasmid to insert into a phage λ genome by means of site specific recombination, responsible for lysogen formation. This is a reversible recombinational insertion which generates phasmids. Phasmid vectors are preferred for the advantage that DNA may be cloned in plasmid vector and recombinant plasmid can be converted to phage which is easier to store, have long shelf life and screening by plaque hybridization gives cleaner results than colony hybridization. M13 is a highly developed phasmid vector being used widely.

Introduction of the recombinant into suitable host: The recombinant vector is now introduced into host cell for its multiplication and expression. The host cells are made competent or permeable by either Calcium chloride treatment for transformation or electrick shock for electroporation. Alternatively, cosmids, phasmids and λ particles are used which then infect *E. coli* cells.

Selection of recombinant clones: After transformation, there is a mixture of host cells which are not transformed or cells transformed with self ligated vector and transformed cells carrying recombinant vectors. So identification of recombinat transformed cells employs the reporter genes of vector.

A reporter gene produces a phenotype which permits either easy selection or quick identification of cells in which it is present. For e.g. genes conferring drug resistance or nutritional deficiencies are selectable markers which allow only cells which possess it to survive under selective conditions.

On the other hand, for easy identification of recombinants, scorable markers are used which produce phenotypes different from those that do not have them such as *lux* gene codes for luciferase which produces phosphorescence, *gus* gene codes for β-galactourinidase which produces blue colour in presence of substrate X-gluc.

The next step is to identify the clone having DNA of interest from a population of transformed cells with recombinants. This is achieved by nucleic acid hybridization methods by colony or plaque hybridization if a specific probe for the DNA insert is available. In cases where inserted gene is expressed, clones synthesizing a foreign protein can be detected immunochemically using specific antibody.

Methods for Gene Transfer

The uptake of foreign DNA or the recombinant DNA by cells is called gene transfer or transformation. Conventionally, the gene transfer necessary for crop improvement is obtained through sexual and vegetative propagation. However, biotechnological approaches like somaclonal variation, protoplast fusion etc. has successfully speeded up the process of generating genetic variation and introgression of foreign genes. The most potential biotechnological approach for transferring recombinant DNA is based on genetic engineering which involves various techniques for gene transfer discussed ahead. The transferred gene is called transgene and the plants that carry these stably integrated transgenes are called transgenic plants. Various gene transfer techniques used are grouped into two broad categories:

- Direct gene transfer
- Agrobacterium mediated.

Direct Gene Transfer is a process where no vector is involved and can be applied to any species or genotype. The methods for direct gene transfer are further classified into two classes:-Physical where usually naked DNA is directly transferred .Therefore, also referred to as DNA mediated gene transfer. The various physical methods for gene transfer are-

Electroporation: is a process where the cells are exposed to electrical impulses of high voltage to reversibly make cell membranes permeable for uptake of DNA. Electroporation has been used extensively for transformation of protoplasts. Recently, transformation of intact plant cells of sugarbeet and rice has been successfully reported. This method is convenient, simple and quick. However, electroporation cannot be applied to all the tissues, cell viability drops due to electric shock. Also, regeneration of plants from protoplasts is still difficult.

Particle bombardment: This is a relatively recent development but is widely used and is effective in introduction of DNA into plant cells. The technique involves coating 1mm diameter particles of tungsten or gold known as microprojectiles with DNA, which are then accelerated to high speed using a pulse of high pressure helium into an evacuated chamber containing the target tissues.

These DNA coated particles penetrate through the cell wall releasing DNA from particles which can express transiently or get integrated into nuclear genome of that cell. With appropriate tissue culture and selection, transgenic plants can be regenerated. Particle bombardment has been used for transformation of monocotyledonous crop plants such as maize, rice, wheat etc.

Agrobacterium mediated Gene Transfer: form the most successful plant transformation system. This is vector based gene transfer method where the vectors are the Ti plasmids of *Agrobacterium tumefaciens* which is a soil inhabiting bacterium responsible for crown gall disease of many plants. The bacterium infects through a wound in the stem of the plant and a tumour develops at the junction of root and stem. The agent responsible for crown gall is Ti plasmid and not the bacterium itself. During infection, a small portion of Ti plasmid DNA, called T-DNA is transferred to the plant cell nucleus, where it becomes covalently inserted into the nuclear DNA, thus getting stably maintained in the genome of transformed cells. T-DNA carries the genes responsible for tumour formation and for synthesis of unusual aminoacid derivatives known as opines (most common are octopine and nopaline).

These opines are used as sole carbon/nitrogen source for inducing Agrobacterium strain. The genes responsible for transfer of T-DNA called virulence genes (*vir* genes) are also contained on the Ti plasmid. Agrobacterium infection requires wounded plant tissue because vir genes are induced by phenolic compounds released by the injured plant cells. The regions of T-DNA absolutely required for its transfer and integration into the plant genome are border regions which are short repeat sequences of 25bp.

Any DNA sequence inserted between the border repeats will be transferred to and integrated into the plant genome. Therefore, Ti-based plasmids are excellent vectors for introducing foreign genes into plants. In order to use them as vectors, the genes responsible for tumour formation must be removed. Ti-based plasmids lacking tumorigenic functions are known as disarmed vectors. These disarmed vectors are still too large to be conveniently used as vectors. Thus, smaller vectors described below have been constructed that are suitable for manipulation *in vitro*.

Cointegration and binary vector system: The cointegration technique is based on *in vivo* recombination of two plasmids. One plasmid carries desirable DNA sequence; the other plasmid contains *vir* genes and the border repeats of T-DNA. The recombination of these plasmids leads to large Ti plasmid which now can be used to transform plants.

The binary vector system uses two separate plasmids: mini-Ti plasmid to supply the disarmed T-DNA and second having *vir* genes. The mini-Ti plasmid bears the gene construct that will be inserted into the plant genome, along with a eukaryotic selectable marker between T-DNA border sequences, so that both genes will be inserted as a unit. This plasmid when placed into an Agrobacterium strain containing a plasmid with virulence functions, the *vir* gene products are able to drive the transfer of T-DNA into plant cells, even though T-DNA is located on a separate DNA molecule.

This is the most frequently used approach as mini Ti plasmids are very easy to manipulate using standard recombinant DNA techniques.

This method is preferred over all other techniques as Agrobacterium is capable of transferring large fragments (<50kb) of DNA very efficiently without much rearrangements. Also, the gene transferred is stably inherited.

2

Biotechnology of Plant Tissue

A mature vascular plant (any plant other than mosses and liverworts), contains several differentiated cell types. These are grouped together in tissues. Some tissues contain only one type of cell. Some consist of several.

Meristematic

The main function of meristematic tissue is mitosis. The cells are small, thin-walled, with no central vacuole and no specialized features.

Meristematic tissue is located in:

- the apical meristems at the growing points of roots and stems.
- the secondary meristems (lateral buds) at the nodes of stems (where branching occurs) , and in some plants,
- meristematic tissue, called the cambium, that is found within mature stems and roots.

The cells produced in the meristems soon become differentiated into one or another of several types.

Protective

Protective tissue covers the surface of leaves and the living cells of roots and stems. Its cells are flattened with their top and bottom surfaces parallel. The upper and lower epidermis of the leaf are examples of protective tissue.

Parenchyma

The cells of parenchyma are large, thin-walled, and usually have a large central vacuole. They are often partially separated from each other. They are usually stuffed with plastids. In areas not exposed to light, colourless plastids predominate and food storage is the main function. The cells of the white potato are parenchyma cells. Where light is present, e.g., in leaves, chloroplasts predominate and photosynthesis is the main function.

Sclerenchyma

The walls of these cells are very thick and built up in a uniform layer around the entire margin of the cell. Often, the protoplasts die after the cell wall is fully formed. Sclerenchyma cells are usually found associated with other cells types and give them mechanical support. Sclerenchyma is found in stems and also in leaf veins. Sclerenchyma also makes up the hard outer covering of seeds and nuts.

Collenchyma

Collenchyma cells have thick walls that are especially thick at their corners. These cells provide mechanical support for the plant. They are most often found in areas that are growing rapidly and need to be strengthened. The petiole ("stalk") of leaves is usually reinforced with collenchyma .

Xylem

Xylem conducts water and dissolved minerals from the roots to all the other parts of the plant. In angiosperms, most of the water travels in the xylem vessels. These are thick-walled tubes that can extend vertically through several feet of xylem tissue. Their diameter may be as large as 0.7 mm. Their walls are thickened with secondary deposits of cellulose and are usually further strengthened by impregnation with lignin. The secondary walls of the xylem vessels are deposited in spirals and rings and are usually perforated by pits.

Xylem vessels arise from individual cylindrical cells oriented end to end. At maturity the end walls of these cells dissolve away and the cytoplasmic contents die. The result is the xylem vessel, a continuous non-living duct. The vessels carry water and some dissolved solutes, such as inorganic ions, up the plant. Xylem also contains tracheids. These are individual cells tapered at each end so the tapered end of one cell overlaps that of the adjacent cell. Like xylem vessels, they have thick, lignified walls and, at maturity, no cytoplasm. Their walls are perforated so that water can flow from one tracheid to the next. The xylem of ferns and conifers contains only tracheids. In woody plants, the older xylem ceases to participate in water transport and simply serves to give strength to the trunk. Wood is xylem. When counting the annual rings of a tree, one is counting rings of xylem.

Phloem

The main components of phloem are :

- sieve elements and
- companion cells.

Sieve elements are so-named because their end walls are perforated. This allows cytoplasmic connections between vertically-stacked cells. The result is a sieve tube that conducts the products of photosynthesis — sugars and amino acids — from the place where they are manufactured (a "source"), e.g., leaves, to the places ("sinks") where they are consumed or stored; such as :

- roots
- growing tips of stems and leaves
- flowers
- fruits, tubers, corms, etc.

Sieve elements have no nucleus and only a sparse collection of other organelles. They depend on the adjacent companion cells for many functions. Companion cells move sugars and amino acids into and out of the sieve elements. In "source" tissue, such as a leaf, the companion cells use transmembrane proteins to take up — by active transport — sugars and amino acids from the cells manufacturing them. Water follows by osmosis. These materials then move into adjacent sieve elements by diffusion through plasmodesmata. The pressure created by osmosis drives the flow of materials through the sieve tubes. In "sink" tissue, the sugars and amino acids leave the sieve tubes by diffusion through plasmodesmata connecting the sieve elements to the cells of their destination. Again, water follows by osmosis where it may:

- leave the plant by transpiration or
- increase the volume of the cells or
- move into the xylem for recycling through the plant.

Plant Tissue Culture: An Overview

Plant tissue culture is a practice used to propagate plants under sterile conditions, often to produce clones of a plant. Different techniques in plant tissue culture may offer certain advantages over traditional methods of propagation, including:

- The production of exact copies of plants that produce particularly good flowers, fruits, or have other desirable traits.
- To quickly produce mature plants.
- The production of multiples of plants in the absence of seeds or necessary pollinators to produce seeds.
- The regeneration of whole plants from plant cells that have been genetically modified.
- The production of plants in sterile containers that allows them to be moved with greatly reduced chances of transmitting diseases, pests, and pathogens.
- The production of plants from seeds that otherwise have very low chances of germinating and growing, i.e.: orchids and nepenthes.
- To clean particular plant of viral and other infections and to quickly multiply these plants as 'cleaned stock' for horticulture and agriculture.

Plant tissue culture relies on the fact that many plant cells have the ability to regenerate a whole plant (totipotency). Single cells, plant cells without cell walls (protoplasts), pieces of leaves, or (less commonly) roots can often be used to generate a new plant on culture media given the required nutrients and plant hormones.

Techniques

Modern plant tissue culture is performed under aseptic conditions under filtered air. Living plant materials from the environment are naturally contaminated on their surfaces (and sometimes interiors) with microorganisms, so surface sterilization of starting materials (explants) in chemical solutions (usually alcohol or bleach) is required. Mercuric chloride is seldom used as a plant sterilant today, as it is dangerous to use, and is difficult to dispose of. Explants are then usually placed on the surface of a solid culture medium, but are sometimes placed directly into a liquid medium, particularly when cell suspension cultures are desired. Solid and liquid media are generally composed of inorganic salts plus a few organic nutrients, vitamins and plant hormones. Solid media are prepared from liquid media with the addition of a gelling agent, usually purified agar. The composition of the medium, particularly the plant hormones and the nitrogen source (nitrate versus ammonium salts or amino acids) have profound effects on the morphology of the tissues that grow from the initial explant.

For example, an excess of auxin will often result in a proliferation of roots, while an excess of cytokinin may yield shoots. A balance of both auxin and cytokinin will often produce an unorganised growth of cells, or callus, but the morphology of the outgrowth will depend on the plant species as well as the medium composition. As cultures grow, pieces are typically sliced off and transferred to new media (subcultured) to allow for growth or to alter the morphology of the culture. The skill and experience of the tissue culturist are important in judging which pieces to culture and which to discard.

As shoots emerge from a culture, they may be sliced off and rooted with auxin to produce plantlets which, when mature, can be transferred to potting soil for further growth in the greenhouse as normal plants.

Choice of Explant

The tissue obtained from the plant to culture is called an explant. Based on work with certain model systems, particularly tobacco, it has often been claimed that a totipotent explant can be grown from any part of the plant. However, this concept has been vitiated in practice. In many species explants of various organs vary in their rates of growth and regeneration, while some do not grow at all. The choice of explant material also determines if the plantlets developed via tissue culture are haploid or diploid. Also the risk of microbial contamination is increased with inappropriate explants. Thus it is very important that an appropriate choice of explant be made prior to tissue culture. The specific differences in the regeneration potential of different organs and explants have various explanations. The significant factors include differences in the stage of the cells in the cell cycle, the availability of or ability to transport endogenous growth regulators, and the metabolic capabilities of the cells. The most commonly used tissue explants are the meristematic ends of the plants like the stem tip, auxiliary bud tip and root tip. These tissues have high rates of cell division and either concentrate or produce required growth regulating substances including auxins and cytokinins.

Some explants, like the root tip, are hard to isolate and are contaminated with soil microflora that become problematic during the tissue culture process. Certain soil microflora can form tight associations with the root systems, or even grow within the root. Soil particles bound to roots are difficult to remove without injury to the roots that then allows microbial attack. These associated microflora will generally overgrow the tissue culture medium before there is significant growth of plant tissue. Aerial (above soil) explants are also rich in undesirable microflora. However, they are more easily removed from the explant by gentle rinsing, and the remainder usually can be killed by surface sterilization. Most of the surface microflora do not form tight associations with the plant tissue. Such associations can usually be found by visual inspection as a mosaic, de-colorization or localized necrosis on the surface of the explant.

An alternative for obtaining uncontaminated explants is to take explants from seedlings which are aseptically grown from surface-sterilized seeds. The hard surface of the seed is less permeable to penetration of harsh surface sterilizing agents, such as hypochlorite, so the acceptable conditions of sterilization used for seeds can be much more stringent than for vegetative tissues. Tissue cultured plants are clones, if the original mother plant used to produce the first explants is susceptible to a pathogen or environmental condition, the entire crop would be susceptible to the same problem, conversely any positive traits would remain within the line also.

Applications

Plant tissue culture is used widely in plant science; it also has a number of commercial applications. Applications include:

- Micropropagation is widely used in forestry and in floriculture. Micropropagation can also be used to conserve rare or endangered plant species.
- A plant breeder may use tissue culture to screen cells rather than plants for advantageous characters, e.g. herbicide resistance/tolerance.
- Large-scale growth of plant cells in liquid culture in bioreactors for production of valuable compounds, like plant-derived secondary metabolites and recombinant proteins used as biopharmaceuticals .
- To cross distantly related species by protoplast fusion and regeneration of the novel hybrid.
- To cross-pollinate distantly related species and then tissue culture the resulting embryo which would otherwise normally die (Embryo Rescue).
- For production of doubled monoploid (dihaploid) plants from haploid cultures to achieve homozygous lines more rapidly in breeding programmes, usually by treatment with colchicine which causes doubling of the chromosome number.

- As a tissue for transformation, followed by either short-term testing of genetic constructs or regeneration of transgenic plants.
- Certain techniques such as meristem tip culture can be used to produce clean plant material from virused stock, such as potatoes and many species of soft fruit.
- micropropagation using meristem and shoot culture to produce large numbers of identical individuals.

Laboratories

Although some growers and nurseries have their own labs for propagating plants by the technique of tissue culture, a number of independent laboratories provide custom propagation services. The Plant Tissue Culture Information Exchange lists many commercial tissue culture labs. Since plant tissue culture is a very labour intensive process, this would be an important factor in determining which plants would be commercially viable to propagate in a laboratory.

Hairy Root Culture

Hairy root culture, also called transformed root culture, is a type of plant tissue culture that is used to study plant metabolic processes or to produce valuable secondary metabolites, often with plant genetic engineering. A naturally occurring soil bacterium *Agrobacterium rhizogenes* that contains root inducing plasmids (also called Ri plasmids) can infect plant roots and cause them to produce a food source for the bacterium, opines, and to grow abnormally.

The abnormal roots are particularly easy to culture in artificial media because hormones are not needed, and they are neoplastic, with indefinite growth. The neoplastic roots produced by *A. rhizogenes* infection have a high growth rate (compared to untransformed adventitious roots), as well as genetic and biochemical stability. Currently the main constraint for commercial utilization of hairy root culture is the development and up-scaling of appropriate (bioreactors) vessels for the delicate and sensitive hairy roots.

Metabolic Studies

Hairy root cultures can be used for phytoremediation, and are particularly valuable for studies of the metabolic processes involved in phytoremediation. Further applications include detailed studies of fundamental molecular, genetic and biochemical aspects of genetic transformation and of hairy root induction.

Genetically Transformed Cultures

The Ri plasmids can be engineered to also contain T-DNA, used for genetic transformation (biotransformation) of the plant cells. The resulting genetically transformed root cultures can produce high levels of secondary metabolites, comparable or even higher than those of intact plants.

Use in Plant Propagation

Hairy root culture can also be used for regeneration of whole plants and for production of artificial seeds.

History of Plant Tissue and Cell Culture

History of Plant Tissue Culture

Plant tissue culture, or the aseptic culture of cells, tissues, organs, and their components under defined physical and chemical conditions in vitro, is an important tool in both basic and applied studies as well as in commercial application. It owes its origin to the ideas of the German scientist, Haberlandt, at the begining of the 20th century. The early studies led to root cultures, embryo cultures, and the first true callus/tissue cultures. The period between the 1940s and the 1960s was marked by the development of new techniques and the improvement of those that were already in use. It was the availability of these techniques that led to the application of tissue culture to five broad areas, namely, cell behaviour (including cytology, nutrition, metabolism, morphogenesis, embryogenesis, and pathology), plant modification and improvement, pathogen-free plants and germplasm storage, clonal propagation, and product (mainly secondary metabolite) formation, starting in the mid-1960s. The 1990s saw continued expansion in the application of the in vitro technologies to an increasing number of plant species. Cell cultures have remained an important tool in the study of basic areas of plant biology and biochemistry and have assumed major significance in studies in molecular biology and agricultural biotechnology.

An important aspect of all biotechnology processes is the culture of either the plant cells or animal cells or microorganisms. The cells in culture can be used for recombinant DNA technology, genetic manipulations etc. Plant cell culture is based on the unique property of the cell-totipotency. Cell-totipotency is the ability of the plant cell to regenerate into whole plant. This property of the plant cells has been exploited to regenerate plant cells under the laboratory conditions using artificial nutrient mediums. With the advances made in genetic engineering, it became possible to introduce foreign genes into cell and tissue culture systems. This led to the development of Genetically Modified (GM) or Transgenic Crops which had improved traits and characteristics.

In the early 19th century, Schleiden and Schwann proposed the concept of the 'cell theory'. In 1902, Gottlieb Haberlandt, the german botanist and regarded as the father of plant tissue culture, first attempted to cultivate the mechanically isolated plant leaf cells on a simple nutrient medium. He did not succeed in achieving the growth and differentiation of the cultured cells, however, he predicted the concept of growth hormones, the use of embryo sac fluids, the cultivation of artificial embryos from somatic cells, etc.

During the period 1902 - 1930, attempts were made to culture the isolated plant organs such as roots and shoot apices (organ culture). Hanning (1904) isolated embryos of some crucifers and successfully grew on mineral salts and sugar solutions. Simon (1908) successfully regenerated a bulky callus, buds, roots from a poplar tree on the surface of medium containing IAA which proliferated cell division. Gautheret, White and Nobecourt (1934-1940) largely contributed to the developments made in plant tissue culture. White (1939) cultured tobacco tumour tissue from the hybrid Nicotiana glauca, and N. Langsdorffii. The period of 1940 - 1970s saw the development of suitable nutrient media to culture plant tissues, embryos, anthers, pollen, cells and protoplasts, and the regeneration of complete plants (in vitro morphogenesis) from cultured tissues and cells. In 1941, van Overbek and co-workers used coconut milk (embryo sac fluid) for embryo development and callus formation in Datura. Steward and Reinert (1959) first discovered somatic embryo production in vitro. Maheswari and Guha (1964) developed the anther culture for the production of haplid plants.

Skoog and Miller (1957) advanced the hypothesis of organogenesis in cultured callus by varying the ratio of auxin and cytokinin in the growth medium. Muir (1953) developed a successful technique for the culture of single isolated cells which is commonly known as paper-raft nurse technique (placing a single cell on filter paper kept on an actively growing nurse tissue). In 1952, the Pfizer Inc., New York (U.S.A) got the US patent and started producing industrially the secondary metabolites of plants. The first commercial production of a natural product shikonin by cell suspension culture was obtained.

In 1980s using Genetic engineering, for the first time, it was possible to introduce foreign genes into cell and tissue culture systems to develop plants with improved characteristics (transgenic crops) which may contribute to the path towards the second green revolution.

Plant Cell and Tissue Culture Techniques

The whole plants can be regenerated virtually from any plant part (referred to as explant) or cells.

Plant tissue culture techniques involve the following steps:

a) Preparation and selection of suitable nutrient media

b) Selection of explants such as shoot tip.

c) Surface sterilization of the explants by disinfectants e.g. sodium or calcium hypochorite solution 0.3- 0.6%) followed by washing the explants with sterile distilled water.

d) Inoculation or Transfer of the explants onto the suitable nutrient medium (sterilized by autoclaving) in culture vessels under sterile conditions (using laminar flow hood).

e) Incubation or Growing the cultures in the growth chamber or plant tissue culture room at optimum physical conditions of light (16 hours of photoperiod), diurnal illumination, temperature (25+/- 20C and relative humidity (50%-60%).

f) Regeneration of plants from cultured plant tissues.

g) Hardening: it is the gradual exposure of plantlets for acclimatization to environmental conditions

h) Transfer of plants to the field conditions following the acclimatization/ hardening of the regenerated plants.

Nutrient Media

The intact plants can make their own food but the *in vitro* culture of plant parts or cells requires a variety of nutrients and suitable physical conditions for their growth. The composition of plant tissue culture medium depends upon the type of plant tissues or cells that are used for culture. No single medium can be used for all types of plants and organs, so the composition of the culture medium for each plant material has to be worked out.

A typical nutrient medium consists of the following components:

a) inorganic nutrients (both micro- and macro-elements - C, H, O, N, P, S, Ca, K, Mg, Fe, Mn, Cu, Zn, B, Mb), The six elements namely nitrogen, phosphorus, potassium, calcium, magnesium and sulfur are the essential macronutrients for tissue culture. The ideal concentration of nitrogen, and potassium is around 25 mmol l-1 while for calcium, phosphorus, sulfur and magnesium, it is in the range of 1-3 mmol l-1.

Among the micronutrients, iron requirement is very critical. Chelated forma of iron and copper are commonly used in culture media.

b) a carbon source and energy source (usually sucrose))- Plant cells and tissues in the culture mediumare heterotrophic and therefore depend on the external carbon for energy. Among the various energy sources, sucrose is the most preferred. During the sterilization of the medium, sucrose gets hydrolysed to glucose and fructose and the plant cells utilize first the glucose and than the fructose. The other carbohydrates such as lactose, maltose, galactose etc. have been used in culture media but with limited success.

c) Organic supplements vitamins (e.g. nicotinic acid, thiamine, pyridoxine and myo-inositol), amino acids (e.g. arginine) The plant cells in culture are able to synthesize vitamins just like natural plants, but in suboptimal quantities which does not support proper growth of cells in culture. Therefore the medium is supplemented with vitamins to achieve good growth of cells. Similarly amino acids are added to the cell cultures to stimulate the cell growth and estabilish the cell lines. Organic acids especially the intermediates of krebs cycle e.g. citrate, malate, succinate, pyruvate also enhances the growth of plant cells. Sometimes antibiotics (e.g. streptomycin, kanamycin) are also added to the medium to prevent the growth of the microorganisms.

d) Growth regulators (e.g. auxins, cytokinins and gibberellins Plant hormones play an important role in growth and differentiation of cultured cells and tissues. The growth hormones included in culture media involve: auxins, cytokinins, and gibberellins. The auxins facilitate the cell division and root differentiation. The cytokinins induce cell division and differentiation and the gibberellins is mainly used to induce plantlet formation from adventive embryos formed in culture. Auxins induce cell division, cell elongation, and formation of callus in cultures. 2, 4-dichlorophenoxy acetic acid is one of the most commonly added auxins in plant cell cultures. Cytokinins, promotes RNA synthesis and stimulate protein and enzyme activities in tissues. Kinetin and benzyl-aminopurine are the most frequently used cytokinins in plant cell cultures.

The ratio of auxins and cytokinins play an important role in the morphogenesis of culture systems. When the ratio of auxins to cytokinins is high, embryogenesis, callus initiation, and root initiation occur. For axillary and shoot proliferation, the ratio of auxins to cytokinins is kept low. Among the gibberellins, gibberellin A3 (GA3) is the most commonly used for tissue culture. GA3 enhances callus growth and induces dwarf plantlets to elongate.

e) Solidifying agents like agar. Generally a gelling agent agar (a polysaccharide obtained from red algae, *Gelidium amansil*) is added to the liquid medium for its solidification. The agar obtained from seaweeds provides solid surface for the growth of cells because in the liquid medium, the tissue will be submerged and die due to lack of oxygen. Cells are grown in suspension medium with out agar but such cultures are aerated regularly either by bubbling sterile air or by gentle agitation. Some other less frequently used solidifying agents are biogel (polyacrlyamide pellets), phytagel, gelrite, and purified agarose.

f) pH - An optimum pH (usually 5.7) is also very important. At pH higher than 7.0 and lower than 4.5, the plant cells stop growing in cultures.

The most extensively used nutrient medium is MS medium.

Major Types of Media

- White's medium - is one of the earliest plant tissue culture media
- MS medium - formulated by Murashige and Skoog (MS) is most widely used for many types of culture systems
- B5 medium - developed by Gamborg for cell suspension and callus cultures and at present it's modified form used for protoplast culture
- N6 medium - formulated by Chu and used for cereal anther culture
- Nitsch's medium developed by Nitsch and Nitsch and used for anther culture.

Preparation of Media

The methodology for media preparation involves preparation of stock solutions (in the range of 10x to 100x concentrations) of highly purified chemicals and demineralized water. The stock solutions are stored in glass or plastic containers and frozen till further requirement. Now a days, plant tissue culture media are commercially prepared, and are available in the market as dry powders. The culture media is usually sterilized in an autoclave at 1210C and 15 psi for 20 minutes. Hormones and other heat sensitive organic compounds are filter sterilized and added to the autoclaved medium.

Maintenance of Aseptic Environment

It is very important to maintain aseptic environment during the in vitro culture of plant cells and tissues. Following are some of the methods adopted for sterilization:

(a) Sterilization of Glassware- The glassware can be sterilized in a hot air oven at 160-1800C for 2-4 hours.

(b) Sterilization of instruments- The metallic instruments are incinerated by dipping them in 75% ethanol followed by flaming and cooling.

(c) Sterilization of nutrient media- The culture media are transferred into glass container, plugged with cotton or sealed with plastic closures and sterilized by autoclaving at 15 psi for 30 min. The autoclaving denatures the vitamins, plant extracts, amino acids and hormones therefore the solution of these compounds are sterilized by using Millipore filter paper with pore size of 0.2 micrometer diameter.

(d) Sterilization of plant materials- The surface of the plant material is made sterile by using disinfectants e.g. sodium hypochlorite, hydrogen peroxide, mercuric chloride, or ethanol. The transfer of sterile plant material on to the nutrient medium is done under the cabinet of laminar airflow.

(e) Sterilization of Culture room and transfer area- the floor and walls of the culture room should be washed with detergent followed by 2% sodium hypochlorite or 95% ethanol. The sterilization can also be done by exposure to UV light. The cabinet of laminar air flow is sterilized by exposing to UV light for 30 min. and 95% ethanol 15 minutes before starting the work.

Types of Culture

Organ Culture

It deals with the culture of the isolated organs (roots) under laboratory conditions *(in vitro)*. Different names are given depending upon the organ used for the culture. For instance the culture of roots, endosperm, ovary, and ovule are called as root culture, endosperm culture, ovary culture etc. It was Skoog (1944), who for the first time suggested that the organogenesis could be chemically controlled. Skoog and Miller (1957) also demonstrated that a high ratio of auxin: cytokinin stimulated the formation of root in tobacco callus, but a low ratio of the same induced shoot information.

Explant Culture

The culture of plant parts (explants) is known as explant culture. The explants can be any part of the plant e.g. the piece of stem, leaf, hypocotyl, etc. The explant cultures are generally used to induce callus or plant regeneration.

Callus Culture

Callus refers to an unorganized mass of cells generally parenchymatous in nature. The unique feature of callus is that the abnormal growth has biological potential to develop normal root, shoots, and embryoids, ultimately forming a plant. Naturally, the callus is formed due to the infection of microorganisms from wounds due to stimulation by endogenous growth hormones, the auxins and cytokinins. However, it has been possible to artificially develop callus by using tissue culture techniques. Auxins are added to culture medium for callus induction but the nature and quantity of auxin added, depends on the nature and source of explant and its genotype besides other factors. Callus cultures can be maintained for prolonged periods by repeated sub-culturing. Callus cultures are used for a) plant regeneration, b) preparation of single cell suspensions and protoplasts, and, c) genetic transformation studies.

Factors Affecting Callus Culture

- the source and the genotype of the explant
- composition of the medium (most commonly used-MS medium)
- temperature (22-280C suitable for callus formation)
- growth regulators e.g. auxins, cytokinins alone or combination of these.

- Age of the plant
- Location of the explant
- Physiology and growth condition of the plant.

Cell Suspension Cultures

Cell suspension is prepared by transferring a fragment of callus to the liquid medium and agitating them aseptically to make the cells free. Single cells can be isolated from either callus or any other part of the plant and cultured in liquid medium using both mechanical and enzymatic methods. Mechanical methods involve grinding of the tissue to a fine suspension in a buffered medium followed by filtration and centrifugation to get rid of cell debris. The enzymatic method uses the enzymes (pectinase or macerozyme) to dissolve the middle lammela between the cells. After the isolation of the cells, they are cultured by batch cultures or continuous cultures. As the medium is liquid in nature, the pieces of callus remain submerged which creates anaerobic conditions. To overcome this problem, the suspension cultures are agitated by a rotary shaker which disperses the cells and expose them to air.

The advantages of cell suspension cultures over the callus culture:

a) The suspension can be pipetted.

b) They are less heterogeneous and cell differentiation is less pronounced.

c) They can be cultured in volumes upto 1,500 litres.

d) They can be subjected to more stringent environmental controls.

e) The manipulations for the production of natural products by feeding precursors, is possible.

Batch cultures are initiated as single cells in 100- 250 ml flasks and are propagated by transferring regularly small aliquots of suspension to a fresh medium. Continuous cultures are maintained in a steady state for long period by draining out the used medium and adding fresh medium. The cell suspension cultures can be used for a) induction of somatic embryos/shoots, b) *in vitro* mutagenesis and mutant selection, c) genetic transformation, d) production of secondary metabolites.

Mass Cell Culture

Plant cells are cultured in specially designed 'plant bioreactors' which essentially do not have a stirrer as plant cells are shear sensitive. In place of stirrer, gas is gently bubbled which provides stirring as well as meet the demand of a higher oxygen supply.

Protoplast Culture

Protoplasts are plant cells without cell wall and can be isolated by using enzymes like cellulases, pectinases) from leaf, seedling, calli, pollen grains, embryo sacs etc. The protoplasts regenerate cell wall, undergo cell division, and form callus.

The callus can also be subcultured. Some of the examples of plant species that have been regenerated from protoplasts are— Cucumis sativus, Capsicum annum, Ipomoea batata, Glycine max, Chrysanthemum sp. These cultures are used for a) various biochemical and metabolic studies, b) fusion of two somatic cells to create somatic hybrids, c) fusion of enucleated and nucleated protoplasts to create Cybrids (cytoplasmic hybrids) and d) genetic manipulation.) drug sensitivity.

Bergmann's Cell Plating Technique (Culture of Single Cells)

In this technique, free cells are suspended in a liquid medium. Equal volumes of liquid and agar media are mixed and rapidly spread in petri dish, which makes the cells evenly distributed in a thin layer after solidification. After sealing the Petri dishes with parafilm, they are examined under the inverted microscope to mark the single cells. Plates are incubated in dark at 250C and cell colonies developing from marked single cells, are used to obtain single cell cultures.

Embryo Culture

Besides, roots, shoots, and pollen, embryos can also be cultured to produce haploid plants. The embryo culture is very useful in conditions where embryo fails to develop due to degeneration of embryonic tissues. It has been used as a routine technique in orchid propagation, in breeding of species showing dormancy.

Embryo Rescue

It has been observed that sometimes, inspite of successful pollination and fertilization, the embryos do not develop. The incompatibility between the embryo and the endosperm or some inherent deficiency also results in the under development of embryo. These immature embryos can be dissected out from the seeds and can be grown artificially on culture medium. These embryos differentiate into shoot, root and plantlets under culture conditions. This technique of growing immature embryo is termed as 'embryo rescue'. This technique is very useful in hybridization, breaking dormancy of certain seeds, and to achieve complete growth of embryo into a plant.

Anther and Pollen Culture (Production of Haploid Plants)

Haploid plants possess a single set of chromosomes (gametophytic no of chromosomes i.e. n) in the sporophyte in contrast to diploids which contain two sets of chromosomes (2n). The existence of haploid plants was reported by Bergner (1921) in Datura stramonium. Tulecke (1951) cultured the pollen grains of Ginko biloba (gymnosperm) and succeeded in inducing the development of haploid callus.

Guha and Maheswari (1964) reported the direct development of haploid embryos and plantlets from microspores of Datura inoxia by the cultures of excised anthers. In 1967, Bourgin and Hitsch obtained the first full haploid plants from Nicotiana tabacum. Haploid plants are very useful in:

a) direct screening of recessive mutation because in diploid or polyploid screening of recessive mutation is impossible.
b) Development of homozygous lines in a short span of time.
c) The generation of exclusive male plants by the process of androgenesis using techniques to double the chromosome numbers.
d) Production of disease resistant plants by introducing disease resistant genes e.g. Barley accession Q 21681 resistant to stem rust, leaf rust, and powdery mildew.
e) Cytogenetic research which includes production of aneuploids, determination of origin and basic number of chromosome numbers, induction of genetic variability.

At present, more than 247 plant species and hybrids belonging to 38 genera and 34 families of dicots and monocots have been regenerated using anther culture technique e.g. rice, wheat, maize, coconut, rubber trees etc.

The Institute of Crop Breeding and Cultivation (China) has developed the high yielding and blast resistant varieties of rice zhonghua No. 8 and zhonghua No. 9 through transfer of desired alien gene.

Androgenesis

In androgenesis, the male gametophyte (the microspore or immature pollen) produces haploid plants by stopping the development of pollen cell into a gamete and forcing it to develop into a haploid plant.

In Vitro Androgenesis

In vitro androgenesis is the formation of sporophyte from the male gametophyte on artificial medium and is most commonly found in family Solanaceae and Poaceae (Graminae).

Direct Androgenesis

In the pollen derived embryogenesis, also called direct androgenesis, the pollen directly acts as a zygote and passes through various embryogenic stages similar to zygotic embryogenesis. Direct androgenesis is very common in many plants of the family Solanaceae and Brassicaceae.

Indirect Androgenesis

The process where the pollen grains instead of normal embryogenesis, divide erratically to develop callus is called indirect androgenesis e.g. in barley, wheat, Coffee etc.

The Technique of anther Culture

The anthers with their filaments are removed from the flower buds after surface sterilization. Under aseptic conditions, the anthers are excised and crushed in 1% acetocarmine to test the stage of pollen development. The anthers in correct stage of development are separated and inoculated on a nutrient medium. The anther cultures are maintained at 280C and alternating photo periods of light (12-18 hrs) and darkness (6-12 hrs). The anthers proliferate and produce callus which forms an embryo and the embryo subsequently develops into a haploid plant.

The Technique of Pollen Culture

The pollens are extracted by pressing and squeezing the anthers with a glass rod against the sides of the beaker. The anther tissue debris is removed by filtering the pollen suspension and large and healthy pollen are washed and collected. These pollen are cultured on a solid or liquid medium and the callus or the embryo forme d is transferred to a suitable medium to produce a haploid plant. The factors that affect androgenesis are the genotype and the physiological state of the donar plants, composition of the culture medium, the method adopted in the pretreatment of anthers to develop in to haploid plants and the stage of the microspore or pollen selected for culture. The haploid are identified by looking at their morphological features or by using genetic markers e.g. a1marker for brown coloured aleurone to detect haploids in maize plant.

Limitations in Haploid Production

a) Due to the low frequency of haploid production the selection is very difficult.

b) Haploids with deleterious traits frequently develop in cultures.

c) It is sometimes difficult to isolate haploids from the culture since the polyploids outgrow haploids.

d) The doubling of haploids (diploidization) does not always lead to the formation of homozygous plant.

e) The embryos derived from haploids often get aborted.

Regeneration Pathways of Plants

The plants can be regenerated by ;

a) organogenesis and

b) somatic embryogenesis.

Organogenesis refers to the formation of organs from the cultured explants. The shoot buds or monopolar structures are formed by manipulating the ratio of cytokinin to auxin in the cultures.

In Somatic embryogenesis, the totipotent cells may undergo embryogenic pathway to form somatic embryos which are grown to regenerate into complete plants. It was demonstrated for the first time in carrots (*Daucus carota*), where bipolar embryos developed from single cells. The somatic embryogenesis is influenced by plant extracts, growth regulators, and by the physiological state of calli.

3

An Introduction to Restoration Genetics

Why is Genetic Diversity Important?

All living organisms carry a genetic blueprint. This is so regardless of whether they are plants, animals, or fungi, whether they are short- or long-lived, and whether they reproduce sexually or clonally. Therefore, to the extent that restoration deals with living organisms, genetics are part of the picture. Although the basic principles underlying restoration genetics may be familiar, to date surprisingly little attention has been devoted to genetic considerations in restoration practice. The purpose of this Restoration Science and Policy Paper is to outline some considerations that restoration designers and managers should be aware of, and to identify more detailed resources that may be useful in practice.

Genotypes Partly Determine Organisms' Physical form and Function

Biologists refer to two basic expressions of variation, the genotype and the phenotype. The *genotype* is the genetic code of an organism; in organisms with nucleated cells (most multi-cellular plants, animals, and fungi), the essential code is found in the nucleus of each cell. Additional genetic codes reside in other components of the cell, such as mitochondria (in animals) and chloroplasts (in plants). An organism's genotype consists of a large number of *genes* (50,000-100,000 in a typical vertebrate), which can be at multiple sites (*loci*) on chromosomes. Genes have a variety of functions, the most important of which is to code for the production of specific amino acids, which are ultimately used to synthesize proteins. In most higher plants and animals (but not fungi, bryophytes, and many marine invertebrates), the adult phase has two copies of each gene, one derived from each parent. When the two copies are the same, the individual is *homozygous* for that gene; if the two copies are different, the individual is *heterozygous* for that gene. The various forms of a gene are referred to as *alleles*; when these forms are identical across a population, the gene is considered *monomorphic*, and if more than one allele exists the gene is considered *polymorphic*.

Phenotype is the expression of these genes as a living organism in a particular environment, and is influenced by environmental context at every level from the cell to the whole organism. It is frequently very difficult to separate variation in an organism's traits with a genetic basis from traits that are phenotypically variable; mistakes are Introduction to restoration genetics. Falk, Knapp, and Guerrant, Society for Ecological Restoration, 17 December 2001.

commonly made both ways (assuming a genetic basis when observed variation is really phenotypic plasticity; underestimating subtle genetic effects on traits that are assumed to be simply "phenotype"). For example, a plant growing in a poor environment might end up having a small and stunted phenotype, despite having "good " genes. This does not mean that genetic variation is not important to the fitness of individuals or populations. Both the environment and the genes will ultimately contribute to restoration success, but steps to ensure an optimal genetic makeup of a population are commonly overlooked. Examples abound in the scientific literature illustrating how genetic composition affects the form and function of organisms. In fact, the recognition of genetic variation among individuals was a primary insight that led to the formulation of evolutionary theory as we know it today (Freeman and Herron 1998).

Genes regulate body size, shape, physiological processes, behavioral traits, reproductive characteristics, tolerance of environmental extremes, dispersal and colonizing ability, the timing of seasonal and annual cycles (*phenology*), disease resistance, and many other traits. Thus, to ignore genetic variation in ecology is to ignore one of the fundamental forces that shape the biology of living organisms. From a restoration perspective, organism and population genetics are fundamental considerations in the design, implementation, and expectations of any project, whether or not explicit consideration is given to the genetic dimension.

Genetic Diversity Helps Organisms Cope with Current Environmental Variability

Organisms exist in environments that vary in time and over space. Such variation is often described in terms of the *natural* or *historic range of variability* (NRV, HRV) in environmental conditions such as weather, disturbance events, resource availability, population sizes of competitors, etc. If a group of organisms (say, a population of species X) were to live in a completely stable physical and biological environment, then a relatively narrow range of phenotypes might be optimally adapted to those conditions. Under these circumstances, Species X would benefit more by maintaining a narrow range of genotypes adapted to prevailing conditions, and allele frequencies might eventually attain equilibrium. By contrast, if the environment is Introduction to restoration genetics.

Examples of traits with a genetic basis for tolerance of environmental variation important in restoration work include tolerance of freezing, drought or inundation, high or low light availability, salinity, heavy metals, soil nutrient deficiencies, and extreme soil pH values in plants; resilience to fluctuating temperature, dissolved oxygen, and nutrient availability in aquatic organisms; and resistance to novel diseases in all groups of organisms (Huenneke 1991). For example, if all individuals in a population are the same genotype with limited drought tolerance, then a single climatic event may destroy the entire population. Plant populations often include individuals with a range of flowering or emergence times. For instance, Great Basin shrub populations include individuals that leaf out and flower over a period of weeks, increasing the likelihood of persistence of the population through periods of unusually early or late growing conditions. Knapp *et al.* (2001) documented flowering periods in a population of individual blue oak trees and found that trees initiated flowering over a period of a month in the spring. Such variability could potentially be adaptive, since it is more likely that at least some trees in the population will flower during warm sunny periods when wind pollination is most successful.

A diverse array of genotypes appears to be especially important in disease resistance. Genetically uniform populations (such as highly inbred crops) are famously vulnerable to diseases and pathogens, which can (and do) decimate populations in which all individuals are equally vulnerable. Such uniformity also predisposes a population to transmit disease from one individual to another: instead of having isolated diseased individuals, nearly every individual may be exposed to disease by direct contact or proximity. More diverse populations are more Introduction to restoration genetics. Falk, Knapp, and Guerrant, Society for Ecological Restoration, 17 December 2001, likely to include individuals resistant to specific diseases; moreover, infected individuals occur at lower density, and thus diseases or pathogens may move more slowly through the population.

Finally, genetic variation is a factor in competition among individuals in real ecological communities. Traits with a genetic basis such as flower size are key factors in competition among individuals. Among animals, behavioral traits may regulate interspecific competition. Since organisms make energetic or life history tradeoffs among traits (for example, allocating energy between growth and reproduction), genetic variability is an important factor in how populations function.

Diversity within Populations Reduces Potentially Deleterious Effects of Breeding Among Close Relatives

In addition to its adaptive value at the population level, genetic variation (or its lack) within individuals can affect their survival and performance. When both copies of a gene (in a diploid organism) are identical (*i.e.*, when an individual is *homozygous* at that gene or locus), the expression of that gene may include traits that are less beneficial to survival or reproduction in particular circumstances. This may lead to physiological or behavioral problems of genetic origin, such as malformed physical structure, poor biochemical balance, improper organ formation and function, altered social behaviour, and susceptibility to disease (Chai 1976).

Homozygosity at key gene loci is a common result of *inbreeding*, which is sexual reproduction among closely related individuals. Small populations and species that do not disperse well (or are constrained by a fragmented landscape from exchanging genes with other populations) can be particularly susceptible to *inbreeding depression*, which is reduced overall survival and reproduction of organisms with low heterozygosity. Inbreeding depression arises from a variety of causes, including expression of unfavorable or deleterious alleles, and often leads to lower survival and birth or Introduction to restoration genetics. Falk, Knapp, and Guerrant, Society for Ecological Restoration, 17 December 2001, reproductive rates. Heterozygosity is not always beneficial, nor does inbreeding always have adverse effects. In some circumstances, a population may be so well adapted to its local circumstances that introducing alleles from other populations actually reduces its performance (*outbreeding depression*) (Templeton 1991). Paradoxically, organisms can experience outbreeding depression for some traits, while suffering the effects of inbreeding depression for other characters. Many organisms (particularly plants) have evolved breeding strategies, such as self-pollination, that allow successful persistence in small populations without apparent short-term effects of inbreeding depression.

Genetic Diversity is the Primary Basis for Adaptation to Future Environmental Uncertainty

Finally, genetic variation holds the key to the ability of populations and species to persist over evolutionary time through changing environments (Freeman and Herron 1998). No organism can predict the future (and evolutionary theory does not require them to), nor can any organism be optimally adapted for all environmental conditions. Nonetheless, the current genetic composition of a species influences how well its members will adapt to future physical and biotic environments.

Consider, for example, the response of a population of plant Species A to a period of rapid climate change. Changes in climate are reflected locally in altered annual rainfall and its seasonal distribution, changes in annual mean maximum and minimum temperatures and their seasonality, changes in prevailing wind direction and speed, and many other factors. These changes cause the zone of suitable climate for Species A to shift geographically. Populations with a diverse array of optima for these climatic conditions (*i.e.*, some individuals that are slightly more cold or heat tolerant, etc.) are more likely to persist, because some individuals will survive the new conditions and reproduce.

If suitable climate zones shift across the landscape, seeds of some individuals may be dispersed into the new location and find good conditions for growing. In this manner, the population can "migrate" across the landscape over generations. By contrast, populations that have a narrower range of genotypes (and are more phenotypically uniform) may simply fail to survive and reproduce at all as conditions become less Introduction to restoration genetics. Falk, Knapp, and Guerrant, Society for Ecological Restoration, 17 December 2001, locally favourable. Such populations are more likely to become *extirpated* (locally extinct), and in extreme cases the entire species may end up at risk of extinction. For example, the Florida Yew (*Torreya taxifolia*) is currently one of the rarest conifer species in North America. But in the early Holocene (10,000 years ago), when conditions in southeastern North America were cooler and wetter than today, the species was probably widespread. For reasons that are not completely understood, *T. taxifolia* failed to migrate northward as climate changed during the Holocene. Today, it is restricted to a few locations in the Apalachicola River Basin in southern Georgia and the Florida panhandle. As the *T. taxifolia* story illustrates, once species are pushed into marginal habitat at the limitations of their physiological tolerance, they may enter an "extinction vortex", a downward cycle of small populations, reduced genetic variability, reduced ability to adapt to novel conditions, leading to further reductions in population size, and so on. Reduced genetic variability is a key step in the extinction vortex.

How is Genetic Diversity Distributed in Natural Populations?

There are genetic differences among individuals within most (but not all) populations of plants and animals. There are also differences among populations across the range of each species. In this section we review some basic patterns of how genetic diversity of species is distributed, or *partitioned.*

Monomorphic and Polymorphic Alleles

By definition, individuals within a given species share some percentage of alleles; otherwise, they would not be considered members of the same species. This shared (or common) portion of the gene pool includes two basic classes of genes. Genes that are monomorphic (*i.e.*, $A = 1$, where A is the number of alleles per locus) within a species are common to all populations, and indeed essentially to all individuals. (Technically, "monomorphic" genes are defined as those for which the frequency of the most common allele is $> 99\%$; thus, some variation may be present even at these loci). The monomorphic proportion of the total genome varies among taxonomic groups, typically in the range of 85% in animals and 50% in plants.

The remainder of loci (on average, 15% in animals and 50% in plants) is Introduction to restoration genetics. Falk, Knapp, and Guerrant, Society for Ecological Restoration, 17 December 2001. polymorphic ($A > 1$), varying among individuals in a population, and among populations within a species. These loci are of most concern to restoration, since polymorphic loci may or may not be adequately represented depending on the design and execution of a restoration program.

Diversity within and Among Populations of a Single Species

The genetic profile of whole populations typically varies from place to place across a species range. These differences may arise as the result of chance occurrences, such as the genetic composition of dispersing individuals that create a new population (*founder effect*), or changes in allele frequencies that result from chance matings in very small populations (*genetic drift*). Differences among populations can also arise systematically, especially if the environment in various places exposes individuals to different optima for survival and reproduction (*fitness*). For these and other reasons, populations often diverge from one another in their genetic composition. Such divergence is especially strong and rapid when there is little *gene flow* between populations (*e.g.*, limited dispersal of seeds or pollen, or limited movement of animals across physiographic barriers). Over evolutionary time, such among-population genetic differences can accumulate and eventually result in the development of a new species (*allopatric speciation*). Indeed, "populations" are defined as much (or more) by patterns of mating and gene flow as by the physical distribution of individuals, although the two are often closely related (Hartl and Clark 1997).

Each species distributes its genetic diversity (one measure of which is the total of all alleles at all loci) in a pattern reflecting both its biology and its history. For example, nearby populations of plants that are pollinated by bees may share many alleles because genes (packaged in pollen grains) can flow easily between sites. Such species may have fewer unique alleles in each population, so populations tend to be genetically similar. By contrast, there may be less gene flow among populations of species that are pollinated by ground-dwelling flightless beetles, or whose heavy fruits fall to the ground in the vicinity of the parent tree. Gene flow can also be obstructed by physical barriers (*i.e.*, topography or habitat that a pollinator, disperser, or migrating individual cannot cross), as well as by disturbance. Introduction to restoration genetics. Falk, Knapp, and Guerrant, Society for Ecological Restoration, 17 December 2001.

Even if variation within a population is low, there may be considerable variability *among* populations. Imagine Species X, a plant that lives in high mountaintop alpine areas, in which every individual on a given mountain is nearly identical genetically to others. In this case, individuals in Population 1 will all be nearly identical to each other; individuals in Population 2 will also be identical to each other, but because populations are very isolated from one another, they may all be very different from all the individuals in Population 1; and so on. Under these circumstances, most of the variation in Species A is *among* populations; within-population variation is low or nonexistent.

Now imagine a contrary example. Species B lives in tallgrass prairie, where suitable habitats are relatively close to one another (or even continuous), and dispersal among populations is common. Here we might find a great deal of variation among individuals in each population (reflecting the benefits of variation discussed above), but because gene flow is high among sites, most populations are similar (that is, most polymorphic alleles are widely distributed). This illustrates a species with a high proportion of variation distributed *within* populations, while among-population variation is relatively small.

A variety of measures are used to quantify the distribution of genetic variation among individuals within populations, and among populations. These measures are the basis for describing how genetic variation is *partitioned* within species. Differences among populations are commonly quantified by the use of one of several statistics, including Wright's inbreeding coefficient (*FST*) and Nei's coefficient of gene variation (*GST*). These indices are functions of how heterozygosity is partitioned within and among populations, based on differences in allele frequencies. Where *pi* is the frequency of the *i*th allele, heterozygosity (*H*) = 1- Ó *pi*2 for {i = 1...n} for populations within a larger sample.

Then, the proportion of total variation that is distributed among populations (*GST*) is *HT- HSHT*, or equivalently, 1- (*HS/HT*), where *HS* and *HT* are the mean heterozygosity within populations and in the entire species respectively. Values for these statistics can often (although not often enough!) be found in the population genetic literature, so restorationists do not have to generate them. Values of *FST* and *GST* vary from 0 to 1.

GST in particular has a number of useful properties: it can be used for one or many loci, mutation rates do not alter the statistic significantly, the exact number of populations need not be specified, and the statistic is relatively responsive to changes in allele frequencies in time. Although they have important conceptual differences, in practice *FST* and *GST* are used in similar fashion as indices of genetic difference among populations (Crow 1986).

Species genetic structure can be modelled in a variety of ways. The simplest is a "finite island" conceptual model, in which alleles can flow between any pair of populations. In this case, the key parameters are the migration rate and *effective population size* (Crow 1986). However, the assumption that any two populations are equally likely to exchange alleles is probably unrealistic. Instead, a "stepping stone" model can be used, in which alleles spread by "stepping" from one population to an adjacent one.

In practical terms, *G* and *F* statistics tell us whether the majority of genetic variation is distributed among or within populations. In species with low *GST* (approaching 0), the majority of variation is found *within populations*; individuals within populations are likely to be genetically different, but each population contains the same complement of alleles in similar frequencies Where *GST* is high (approaching 1), individuals within a population are relatively similar but *populations* are significantly different. Most species fall somewhere in between these extremes.

The distribution of genetic variation within and among species is strongly linked to life-history traits, particularly dispersal and *reproductive mode*. Species that disperse genes (in plants, this includes both pollen and seeds) widely and frequently will tend, other things being equal, to have lower *GST* (*i.e.*, populations will be more similar). Even a moderate rate of gene movement among populations (one individual every a few generations) can "link" the gene pools of two populations. Mutation and drift (chance selection of genotypes, especially likely in small populations) can also lead to changes in allele frequencies, although in general these forces are believed to act more slowly than dispersal and selection.

How is Genetic Variation Detected and Measured?

A variety of methods exist for the assessment of genetic variation. Traditionally, genetic variation was inferred by the shape (*morphology*) or growth responses of organisms. Such approaches often use *common garden experiments* or *reciprocal transplants* to distinguish genotypic and phenotypic variation. In this approach, individuals that appear phenotypically different (or that grow in different environments) are placed in a common environment. This eliminates the environmental component of variation; presumably, the remaining variation has a genetic basis. variation. While these methods can be useful, they measure higher-order effects of genetic variation (*i.e.*, expression in the whole organism), and hence address genetic variation only indirectly. Moreover, the confounding effects of phenotypic variation are often difficult to separate from underlying genetic differences among organisms. In practical terms, it is often not possible to know whether organisms that look different are actually different genetically, or whether their differing phenotypes reflect the influences of environment or chance in development.

Since the 1960's the most widely applied method of estimating genetic differences among individuals is *enzyme electrophoresis* (also referred to as *starch gel electrophoresis*, including both *allozyme* and *isozyme* analysis). Electrophoresis detects variation in amino acids that are early products of gene *translation* (the process by which a cell translates a copy of its genetic code in messenger-RNA into amino acids). Since there may be multiple biochemical pathways leading to the synthesis of a single amino acid or enzyme, electrophoresis is relatively conservative in its estimates of genetic variation among individuals. Enzyme electrophoresis remains the most widely used method of estimating genetic variation, and literally thousands of studies have been conducted on a wide range of organisms. Electrophoretic analysis permits individuals to be distinguished from one another based on enzyme variation. These differences among individuals can be compiled to generate statistics about the degree of variation within and among populations (§2.b), a key consideration in restoration ecology.

In the past decade, powerful methods have developed and emerged to detect variation in DNA itself, not simply its products. While a
Introduction to restoration genetics. Falk, Knapp, and Guerrant, Society for Ecological Restoration, 17 December 2001.

bewildering variety of tools and techniques may be encountered in the literature, they all have in common the ability to reconstruct the sequence of nucleotides in the DNA molecule, arguably the most basic level of genetic variation. The conceptually simplest (although experimentally most laborious) technique is *gene sequencing* — that is, the exact reconstruction of the complete nucleotide sequence. Part of the difficulty in interpreting DNA sequences is that "genes" in a functional sense may be distributed over several sections of a chromosome; there are also large sections of all chromosomes that do not code for any known function, or that are redundant or *regulatory* (influence the transcription, translation, or function of other genes). Thus, unlike enzyme analysis, much of the information in a DNA sequence may be of little relevance to understanding ecologically significant genetic variation. Since in theory every sexually-produced individual is likely to be unique, DNA sequences can be almost too variable to detect the kinds of patterns required for restoration work. However, new analytical techniques allow DNA sequence variation to be interpreted in meaningful ways. Complete gene sequences have been complied for relatively few organisms, and generally only a few individuals each, although these numbers are increasing all the time.

It is important to note that isozymes and genetic markers may not reflect traits under strong natural selection. For example, Knapp and Rice ((1998) evaluated patterns of variation in a native grass using both quantitative traits and isozymes, and compared these to geographic distance and climate. They found overall that isozymes did not reflect an adaptation to climate, whereas the quantitative traits did. In a restoration context, this suggests that climate may in some cases serves as more useful guide for collecting and reintroduction zones that genetic markers *per se*.

How do Genetic Questions Enter into Restoration?

The genetic principles outlined above have a direct bearing on the practice of ecological restoration. In this section we summarize how restoration practitioners and researchers can (and should) take population genetics into account.

Accuracy and Functionality of Restored Populations

Sophisticated restoration practitioners recognize that the starting pool of genetic variation to be used is a critical element in design and implementation. However, opinions abound on how to select a suitable source pool for a restoration project.

The current debate often confuses two related, but distinct issues: the *accuracy* ("authenticity") of a restoration project, and its *functionality* (Clewell, 2000). A reintroduction project is genetically *accurate* ("authentic") if it replicates the original gene pool of the population it replaces. If the original population has been destroyed, then perfect accuracy is strictly speaking impossible, because some alleles were probably unique to that population (*i.e.*, GST < 1). A restoration can be accurate and function poorly; it can also be inaccurate but functional. Since no reintroduction is perfectly accurate genetically, the question then becomes: "How close is close enough?" To this there can be no absolute ecological answer, since not all variation is of adaptive significance. Moreover, variation among populations is a continuous variable: two populations can differ by one allele or thousands. If the motivation is an ethical commitment to fidelity to the historic distribution of genotypes, the line can be drawn anywhere. The best practice is to estimate and report the degree of genetic accuracy based on available information, so that others studying a project can take authenticity into account.

How well a reintroduced population functions is a very different matter. Here the concern is not so much whether specific genotypes have been re-established, but rather how well a restored population will work in terms of persistence, resilience, and stability. Since many adaptive traits have a genetic basis, reintroduction material that performs (survives, grows, reproduces) well may come from a site with similar ecological conditions, but not necessarily close geographically. Moreover, if functionality is the main objective, then a range of genotypes can be introduced, allowing selection (differential survival and reproduction) to sort out the best for the site. If this approach is taken, there may be considerable attrition, which can be accounted for in initial sampling.

Geographic Location of Source Material for (Re) Introduction

Perhaps no single genetic issue is more intensely debated in restoration circles than the location of source populations: "Where should material for restoration come from?" Interest in population genetic variation often arises in the context of asking whence source material should come. In other words the restorationist is concerned not only with the degree of variability but its particular geographic distribution.

The most common approach is to specify a geographic (and often elevation) range, within which source material can be confidently collected. This is essentially a "space-for-genotype" substitution: we expect that populations near one another, and growing in similar conditions, will be more similar genetically due both to ecotypic variation and the effects of gene flow. Geographic distance is a reasonable first approximation of genetic distance, assuming that genetic diversity most likely obeys the "distance decay of similarity" in broad outline. In population genetic terms, this favours the "stepping stone" model of gene flow. On public lands in the western United States, commonly-applied guidelines suggest that material for outplanting be collected from within 1,000 ft. elevation bands and 100 miles lateral distance of the planting area (elevation bands vary from 500-3,000 ft for different agencies). As noted above climatic zones may also offer a better first approximation of genetic distance between populations than simple geographic distance. Unfortunately, there are no simple distance rules that apply equally to all species, because species vary in gene flow among populations, population size, and the resulting distribution of diversity (GST). For some species (*e.g.*, self-fertilizing plants in small, isolated patches of habitat, or fishes in isolated stream reaches), each site may reflect a unique local adaptation, and the geographic range of suitable genotypes can be very small (a few km2). Other species (for example, species with wind dispersed pollen and seeds) with higher rates of gene flow, and those with larger effective populations, are probably less differentiated over the landscape and can be collected from much wider ranges. Most species have dispersal curves that are leptokurtic (*i.e.*, with long tails), meaning that a few seeds or offspring in each generation may travel far beyond the majority; while few in number, these long-range dispersers may play a critical role in helping species to adjust their ranges in periods of climate and vegetation change. For each species, the Introduction to restoration genetics. Falk, Knapp, and Guerrant, Society for Ecological Restoration, 17 December 2001.

Restorationist must ask: "How widely does this species disperse its genes under normal conditions, and what factors (distance, geographic barriers, habitat types) influence where genes can spread?" This uncertainty is reflected in the wide range of recommendations for suitable collecting zones, which range over more than three orders of magnitude from 100 meters to 100 miles.

The issue of geographic range for source material cannot be separated from the distinction made earlier between historical accuracy and functionality. If historical accuracy (authenticity) *sui generis* is the primary concern, then the range of potential collecting sites is governed most strongly by historical patterns of dispersal, and the collecting radius will be very small. If population function is the primary concern, on the other hand, then the net may be cast over a wider area (although still focusing on similar ecological communities). This is because populations that are widely separated in space may nonetheless occupy similar ecological settings, and by selection may have developed genotypes that are similar in key ecological traits. If the main goal is population function, then it is probably more important to derive source material from large, diverse populations than from small, genetically depauperate sites, even if the latter are closer geographically. Of course, it is always possible to mix genotypes from different populations, and to let selection sort out the variation: this is, after all, exactly what nature does. Mixing of genotypes may be particularly suitable if existing "populations" are in fact isolated and reduced remnants of formerly widespread and interconnected groups. Re-combining fragmented populations genetically is appropriate as the geographic scale is ecologically realistic, and should not be done indiscriminately. Mixing too broadly (*i.e.*, combining genotypes from very diverse ecological settings) can result in too many individuals that are poorly adapted for the new given environment (high genetic load). Some geneticists advocate the use of regional mixtures — composite collections of genotypes, all of which are at least moderately well adapted to a given environment. before the population dies out.

There are valid arguments on both sides of the authentic *vs.* functionality debate. On one hand, there is little question that the gene pools of many remnant native populations have been seriously eroded, so that what occurs today is a small remnant of the original Introduction to restoration genetics. Falk, Knapp, and Guerrant, Society for Ecological Restoration, 17 December 2001.

Diversity. Small gene pools are more prone to inbreeding, as well as random genetic change from drift. Populations that formerly exchanged genes regularly may have also become genetically isolated by habitat fragmentation. In such cases, a good argument can be made to bring together genetic material from several nearby populations, in effect replicating the natural (but now disrupted) processes of gene flow. In addition, some restoration sites may be so heavily disturbed (*i.e.* mine tailing reclamation areas) that the most "local" population may no longer be the one best adapted to the growing environment. In such cases, a wide diversity of genotypes can increase the chances that at least some plants will have the genetic composition to survive..

It is possible, however, that genotypes from outside the apparent current range may perform too well — that is, swamp the local genotype. If remnant native populations of the species being introduced already exist at a site, additional genetic considerations reinforce the importance of the scale of collection. Introduced populations may hybridize with the existing native population, introducing new genes (genetic pollution) and potentially negatively affecting genetic integrity. If a few poorly adapted individuals make it into the existing native population, it might be argued that natural selection will eventually remove the deleterious genes. However, the issue is partly one of numbers; with commercial seed production of many native species, restorationists now have the tools to dump large volumes of seed into ecosystems. If the number of poorly adapted, non-local propagules is large in relation to the number of local native types, the chance of matings between the well adapted and the poorly adapted plants will increase (especially in cross-pollinating species), thereby potentially swamping the native population and diluting the local genes. Progeny from such matings may additionally experience outbreeding depression (*i.e.* poor survival and growth in relation to the parents). Carefully chosen introductions, with genotypes similar to the existing native population, can avoid these negative impacts.

Take-home messages: Since no single rule applies to all situations, the most important recommendation is to use explicit and reasoned criteria in selecting source populations. To summarize the issues presented here: Introduction to restoration genetics. Falk, Knapp, and Guerrant, Society for Ecological Restoration, 17 December 2001.

1. Species vary in their dispersal rates and distances, and hence in the degree of genetic differentiation among populations. These differences are often closely correlated with life-history attributes particular to each species that affect rates and patterns of gene flow. Therefore, guidelines are inherently species-specific.
2. The restorationist must decide if historical accuracy or functionality is the primary objective. True historical accuracy can be difficult to determine if existing populations are remnants of a formerly widespread range. Moreover, most natural populations experience a degree of genetic change over time and space. However, even where the emphasis is on functionality, historical reference conditions are essential to "anchor" restoration work within a natural range of variability.
3. Geographic distance (*i.e.*, the "space-for-genotype" substitution) is a reasonable, but crude, substitute for patterns of gene flow among populations. However, in a disrupted landscape the most local remnant populations may be genetically reduced, and may not have genotypes that correspond to the conditions under restoration. Moreover, if populations are strongly selected to local habitat conditions, then habitat similarity may outweigh distance as a selection criterion. Local and regional climatic and soil zones can be useful criteria for obtaining material for reintroduction, since these will often represent plants adapted to the conditions at the restoration site.

4. As a source of material for reintroduction, large, genetically diverse source populations are generally preferable to small, limited populations even when the latter are closer to the restoration site. It may be preferable to combine material from several suitable sites, to capture a wider array of genotypes that can succeed in the new location. Such "regional mixtures" should draw on populations within a similar ecological zone (climate, soil) to the restoration site, to avoid too large a proportion of individuals that may not be well adapted to the new conditions.
5. Small local populations can be "swamped" by non-native genotypes if a species already exists at a restoration site. If existing populations are to be Introduction to restoration genetics. Falk, Knapp, and Guerrant, Society for Ecological Restoration, 17 December 2001, augmented, the number of individuals introduced from other locations should not be so large as to overwhelm the local gene pool, particularly if there is evidence of local adaptation.

Sampling the Diversity of Source Populations

The initial gene pool for a restoration project is inevitably limited by the diversity of the original sample. While other genes may enter the project area over time (for instance, by migration of individuals, dispersal of gametes, or additional reintroduction measures), the starting pool of genetic diversity will govern the performance of a reintroduced population for a long time. Hence, the diversity of the original source collection is a critical consideration in restoration.

A variety of guidelines have been developed for sampling wild populations of plants and animals for breeding and reintroduction (Centre for Plant Conservation 1991; Guerrant 1992; BGCI 1994; Guerrant 1996; IUCN/SSC 1998). These guidelines address four fundamental sampling issues:

1. How many populations will be sampled to create the source pool?
2. How many individuals will be sampled from each population?
3. Will samples be collected all at once, or over a period of years?
4. What is the probability of a collected sample surviving to establishment?

Number of populations: In most species, the cumulative amount of genetic variation captured increases as successive populations are added to the sample. However, since populations have some measure of similarity (0 < GST < 1), each additional population added to a sample collects some alleles that are new to the sample, and some that are already present from previous samples. As the number of populations sampled increases, the *marginal diversity rate* decreases (that is, fewer and fewer novel alleles are captured), and the *cumulative diversity function* approaches an asymptote. For a pool of populations sampled at random, there comes a point at which further sampling across populations provides little or no additional genetic benefit.

The number of populations at which this occurs is related strongly to the measure of differentiation among populations, GST, which we have already encountered. When GST is high, populations are markedly different from one another, and more should be sampled Introduction to restoration genetics. Falk, Knapp, and Guerrant, Society for Ecological Restoration, 17 December 2001, to capture the maximum total diversity. When GST is low, most populations are similar, and sampling from only a few will capture most of the diversity that exists(provided that the statistic was generated using populations across the full range of the species).. Of course, it is always worth remembering that a significant fraction of most genomes is monomorphic, and is therefore captured in the first population, if not the first individual. There can be no hard and fast rules for the number of populations to be sampled across all taxa. In a recent survey of 13 plant species, the sample for some species was saturated for common isozyme alleles after just 3-5 populations (20% of the total), while in other species the cumulative diversity curve was still rising after all populations (as many as 15-30) were included in the sample (recall, however, that isozyme loci may be less variable among populations than other loci under strong selection). (Falk and Gibbs *In prep.*). Rare or uncommon alleles accumulate in the sample at a slower rate than do common alleles, so if it is suspected that important adaptations occur at low frequency (for instance, because of recent environmental change), then it is advisable to sample a larger number of sites. Further research is ongoing in this area; in the meantime, GST or some comparable measure of population differentiation is a suitable guide for the sampling strategy.

One exception to this general null model is for populations that are strongly differentiated along habitat lines (*e.g. ecotypes*). If the reintroduction area is unusual habitat for a species, then it is worth sampling any populations that occur in similar conditions, to increase the likelihood of capturing alleles that offer adaptive benefit.

Number of individuals to sample within populations. The underlying theoretical basis for sampling multiple individuals within a population is that populations are rarely truly *panmictic* (that is, with completely randomized breeding). In plants, a large proportion of mating occurs between neighboring individuals, even when pollination occurs via an animal vector. In animal populations, a wide range of behavioral adaptations exists that commonly concentrates breeding success in a few individuals at any given time. The result is that populations are not genetically homogeneous; to capture their genetic diversity adequately, multiple individuals need to be sampled. Again, a number of general guidelines have been released on this topic, and again, there can be no hard and fast "magic numbers" that apply across taxa. Brown and Briggs 1991). Plant studies evaluated by the Centre for Plant Conservation and its colleague institutions suggest that anywhere from 10-50 individuals should be sampled per population, with the understanding that there is bound to be considerable redundancy among these samples. Sampling of propagules (seeds, vegetative shoots) has less impact on the source population, although there is often lower success associated with these life stages compared to outplanting established plants.

Number of Years During which Samples will be Collected

For many species, an adequate sample can be collected in a single year. However, for many others, collecting may need to be distributed over a number of years. Reasons for multiple-year collecting include:

- Variable reproductive output (masting), where large numbers of individuals are reproductive or large numbers of offspring are produced only in certain years;
- Small populations or species with low reproductive rates, where it is important to avoid depleting the pool of new individuals;
- Annual organisms, for which gene pools may be significantly different from year to year, either by selection or by chance; and
- Limited capacity to handle (store, propagate, release) collected material.

Probability of a Collected Sample Surviving to Establishment.

In the end, what counts in a reintroduction is the number of individuals in the new population as well as their diversity (Menges 1991). However, less than 100% of the samples (seeds, cutting, eggs, adults) collected in the field will survive to establishment. Attrition occurs at every step along the way: during the collecting process, transportation, storage, propagation/curation, and outplanting/release. High initial mortality rates are often observed in reintroduced populations, often continuing for several years.

A simple calculation can help to account for attrition of collected samples. Let *Ps* represent the survival probability of a collected sample *s*, and *N* the number of individuals desired in the final restored population. Then we account for attrition during the reintroduction process by collecting *PsN* samples in the field. For example, suppose Introduction to restoration genetics. Falk, Knapp, and Guerrant, Society for Ecological Restoration, 17 December 2001. that trees dug for transplant have an ultimate survival rate of 40% from initial collecting to the third year of a reintroduction project, and that we are trying to establish a population of 50 trees. Correcting for attrition, an initial collection of (50/.40), or 125 individuals, will give us a good chance of ending up with the final population size we wish.

Genetic Diversity within Reintroduced Populations

As discussed above, breeding among closely related individuals can lead to expression of unfavourable traits that compromise an organism's ability to survive and reproduce. Individuals resulting from inbreeding in populations where this did not occur naturally from highly inbred lines can have stunted growth, altered behaviour, deformed morphology, poor physiological function; such individuals are also more likely to be reproductively sterile. Hence, the genetic origin of individuals used in a restoration project can influence whether a project succeeds or fails. If source collections are made from diverse, naturally occurring populations, the probability of breeding among close relatives is reduced. A more diverse population may include individuals that can tolerate a wider range of conditions. If, on the other hand, all individuals in a restored population are genetically similar, and the total population size is not large (fewer than 100 individuals), the probability is great that their offspring will be homozygous at some loci, and thus have reduced fitness. A genetically narrow population may be able to survive only in a narrow range of conditions. While genetic diversity is not always associated with *ecological amplitude*, in general the correlation is positive (Huenneke 1991).

In addition to affecting how well a restored population succeeds in current environmental conditions, genetic diversity may affect performance over time. Important ecological traits, such as tolerance of disturbance or climatic extremes often have a strong genetic basis. Such traits need to be considered not only in terms of conditions at the time of reintroduction, but in light of the natural range of variability over many years or generations. If the goal of restoration is to establish self-sustaining populations (Pavlik 1996), then the range of conditions to which the population and its offspring will be exposed is a vital consideration. Genetically uniform populations may do well one year when conditions are favourable for their genotype, and then fail the next, even though the range of environmental variation is within normal limits. Such failures are even more Introduction to restoration genetics. Falk, Knapp, and Guerrant, Society for Ecological Restoration, 17 December 2001, likely in the case of less common extreme events (Gaines and Denny 1993). Recalling from § 1.b and 1.d that environmental conditions vary within an envelope of natural variation, it is safe to assume that most reintroduced populations will be exposed to a wide range of conditions over time. Hence, the genetic diversity of reintroduced populations is a non-trivial consideration for their long-term persistence (Montalvo et al. 1997).

It is not uncommon in restoration work for large numbers of individuals to be released or outplanted, a large proportion of which fail to survive beyond 2-3 years. This means that restoration releases are subject both to the *founder effect* (the result of the initially limited gene pool) and then further *genetic bottlenecks* as effective population size diminishes due to mortality. If the initial outplanting or release is itself genetically uniform (*e.g.*, hundreds or thousands of clonally-produced plants), the resulting gene pool in the field can be quite narrow. Such apparently large populations are, in genetic terms, very small. Of course, genetically narrow populations occur in nature, often resulting from similar forces (founder events and bottlenecks). For example, aspen (*Populus tremuloides*) and many species in the Iridaceae (*Iris, Lilium*) are often found in large clumps of many *ramets* of very few *genets*, or even just one. Natural populations of species where sexual (as opposed to vegetative or parthenogenic) reproduction dominates are less likely to be genetically homogeneous (Meffe and Carroll 1994).

The effects of genetic homogeneity in a reintroduced population may not be immediately evident, but over a period of years the population may have lower rates of growth, survival and reproduction, and may persist less successfully through periods of natural environmental variability.

What you can do: Restorationists should be aware of the level of genetic diversity that they are working with. Above all, restoration planners and managers should understand how the plants and animals they use were generated. Ask your nursery or whoever collected and propagated your seed source (or your breeder, in the case of animals) what methods were used, and what steps were taken to assure the presence of a suitably wide range of genotypes. While clonally-produced flats of thousands of identical plants may appear to offer short-term advantages (as they do for agricultural crops) of Introduction to restoration genetics. Falk, Knapp, and Guerrant, Society for Ecological Restoration, 17 December 2001, predictable response to current growing conditions, in the long run (again as with crops) such populations may be less likely to persist and succeed in the face of disease, competition, and climate variability. Of course, knowing the methods by which individuals or material was produced provides genetic information only by inference; it is unfortunately not common to have good genetic data for reintroduction efforts.

Genetic Diversity Among Reintroduced Populations

The same principles apply to diversity among population diversity. Most species have a degree of differentiation among populations (*i.e.*, GST > 0). Population geneticists debate whether these differences have adaptive value in any given case – that is, whether they are the result of selection – or whether they represent the effects of sampling error (founder effects, drift, mutation). Nonetheless, a reasonable precautionary position is to mimic the degree of genetic difference found in natural populations of a species when (re)introducing populations to the landscape.

Most species, however, have little or no published data on their among-population genetic diversity. The few surveys that have been done are invaluable sources of information, and restorationists should read and cite them whenever possible. In some cases studies of *congeners* can be used, although with the caveat that species that are closely related phylogenetically are not always similar ecologically.

Resources

A vast literature exists on genetic diversity, its distribution in space and time, and its adaptive and ecological significance. A number of excellent texts on population genetics provide greater depth on a number of topics addressed briefly in this paper. Several of the growing numbers of papers on the topic of restoration genetics are listed in the literature cited section of this paper. Conservation genetics is still is relatively small field, and restoration genetics is in its infancy. Restoration practitioners and scientists can contribute to the accumulation of knowledge in this area by treating your restoration projects as experiments, and publishing your results.

The Society for Ecological Restoration maintains an on-line *Directory of Restoration Expertise* at www.ser.org. The IUCN also has published a list of Introduction to restoration genetics. Falk, Knapp, and Guerrant, Society for Ecological Restoration, 17 December 2001, reintroduction practitioners (Soorae and Seddon 1998).

How Genes Work

People have known for many years that living things inherit traits from their parents. That common-sense observation led to agriculture, the purposeful breeding and cultivation of animals and plants for desirable characteristics. Firming up the details took quite some time, though. Researchers did not understand exactly how traits were passed to the next generation until the middle of the 20th century. Now it is clear that genes are what carry our traits through generations and that genes are made of deoxyribonucleic acid (DNA). But genes themselves don't do the actual work. Rather they serve as instruction books for making functional molecules such as ribonucleic acid (RNA) and proteins, which perform the chemical reactions in our bodies. Proteins do many other things, too. They provide the body's main building materials, forming the cell's architecture and structural components. But one thing proteins can't do is make copies of themselves. When a cell needs more proteins, it uses the manufacturing instructions coded in DNA.

The DNA code of a gene—the sequence of its individual DNA building blocks, labeled A (adenine), T (thymine), C (cytosine), and G (guanine) and collectively called nucleotides— spells out the exact order of a protein's building blocks, amino acids. Occasionally, there is a kind of typographical error in a gene's DNA sequence. This mistake—which can be a change, gap, or duplication—is called a mutation. A mutation can cause a gene to encode a protein that works incorrectly or that doesn't work at all. Sometimes, the error means that no protein is made. But not all DNA changes are harmful. Some mutations have no effect, and others produce new versions of proteins that may give a survival advantage to the organisms that have them. Over time, mutations supply the raw material from which new life forms evolve.

What are Proteins and what do they Do?

Proteins are large, complex molecules that play many critical roles in the body. They do most of the work in cells and are required for the structure, function, and regulation of the body's tissues and organs. Proteins are made up of hundreds or thousands of smaller units called amino acids, which are attached to one another in long chains. There are 20 different types of amino acids that can be combined to make a protein. The sequence of amino acids determines each protein's unique 3-dimensional structure and its specific function.

How do Genes Direct the Production of Proteins?

Most genes contain the information needed to make functional molecules called proteins. (A few genes produce other molecules that help the cell assemble proteins.) The journey from gene to protein is complex and tightly controlled within each cell. It consists of two major steps: transcription and translation. Together, transcription and translation are known as gene expression. During the process of transcription, the information stored in a gene's DNA is transferred to a similar molecule called RNA (ribonucleic acid) in the cell nucleus. Both RNA and DNA are made up of a chain of nucleotide bases, but they have slightly different chemical properties. The type of RNA that contains the information for making a protein is called messenger RNA (mRNA) because it carries the information, or message, from the DNA out of the nucleus into the cytoplasm.

Translation, the second step in getting from a gene to a protein, takes place in the cytoplasm. The mRNA interacts with a specialized complex called a ribosome, which "reads" the sequence of mRNA bases. Each sequence of three bases, called a codon, usually codes for one particular amino acid. (Amino acids are the building blocks of proteins.) A type of RNA called transfer RNA (tRNA) assembles the protein, one amino acid at a time. Protein assembly continues until the ribosome encounters a "stop" codon (a sequence of three bases that does not code for an amino acid).

Can Genes be Turned on and Off in Cells?

Each cell expresses, or turns on, only a fraction of its genes. The rest of the genes are repressed, or turned off. The process of turning genes on and off is known as gene regulation. Gene regulation is an important part of normal development. Genes are turned on and off in different patterns during development to make a brain cell look and act different from a liver cell or a muscle cell, for example. Gene regulation also allows cells to react quickly to changes in their environments. Although we know that the regulation of genes is critical for life, this complex process is not yet fully understood.

Gene regulation can occur at any point during gene expression, but most commonly occurs at the level of transcription (when the information in a gene's DNA is transferred to mRNA). Signals from the environment or from other cells activate proteins called transcription factors. These proteins bind to regulatory regions of a gene and increase or decrease the level of transcription. By controlling the level of transcription, this process can determine the amount of protein product that is made by a gene at any given time.

How do Cells Divide?

There are two types of cell division: mitosis and meiosis. Most of the time when people refer to "cell division," they mean mitosis, the process of making new body cells. Meiosis is the type of cell division that creates egg and sperm cells. Mitosis is a fundamental process for life. During mitosis, a cell duplicates all of its contents, including its chromosomes, and splits to form two identical daughter cells. Because this process is so critical, the steps of mitosis are carefully controlled by a number of genes. When mitosis is not regulated correctly, health problems such as cancer can result.

The other type of cell division, meiosis, ensures that humans have the same number of chromosomes in each generation. It is a two-step process that reduces the chromosome number by half—from 46 to 23—to form sperm and egg cells. When the sperm and egg cells unite at conception, each contributes 23 chromosomes so the resulting embryo will have the usual 46. Meiosis also allows genetic variation through a process of DNA shuffling while the cells are dividing.

How do Genes Control the growth and Division of Cells?

A variety of genes are involved in the control of cell growth and division. The cell cycle is the cell's way of replicating itself in an organized, step-by-step fashion. Tight regulation of this process ensures that a dividing cell's DNA is copied properly, any errors in the DNA are repaired, and each daughter cell receives a full set of chromosomes. The cycle has checkpoints (also called restriction points), which allow certain genes to check for mistakes and halt the cycle for repairs if something goes wrong.

If a cell has an error in its DNA that cannot be repaired, it may undergo programmed cell death (apoptosis) (illustration). Apoptosis is a common process throughout life that helps the body get rid of cells it doesn't need. Cells that undergo apoptosis break apart and are recycled by a type of white blood cell called a macrophage (illustration). Apoptosis protects the body by removing genetically damaged cells that could lead to cancer, and it plays an important role in the development of the embryo and the maintenance of adult tissues.

Cancer results from a disruption of the normal regulation of the cell cycle. When the cycle proceeds without control, cells can divide without order and accumulate genetic defects that can lead to a cancerous tumor (illustration).

How do Geneticists Indicate the Location of a Gene?

Geneticists use maps to describe the location of a particular gene on a chromosome. One type of map uses the cytogenetic location to describe a gene's position. The cytogenetic location is based on a distinctive pattern of bands created when chromosomes are stained with certain chemicals. Another type of map uses the molecular location, a precise description of a gene's position on a chromosome. The molecular location is based on the sequence of DNA building blocks (base pairs) that make up the chromosome.

Cytogenetic Location

Geneticists use a standardized way of describing a gene's cytogenetic location. In most cases, the location describes the position of a particular band on a stained chromosome:

17q12

It can also be written as a range of bands, if less is known about the exact location:

17q12-q21

The combination of numbers and letters provide a gene's "address" on a chromosome. This address is made up of several parts:

- The chromosome on which the gene can be found. The first number or letter used to describe a gene's location represents the chromosome. Chromosomes 1 through 22 (the autosomes) are designated by their chromosome number. The sex chromosomes are designated by X or Y.
- The arm of the chromosome. Each chromosome is divided into two sections (arms) based on the location of a narrowing

(constriction) called the centromere. By convention, the shorter arm is called p, and the longer arm is called q. The chromosome arm is the second part of the gene's address. For example, 5q is the long arm of chromosome 5, and Xp is the short arm of the X chromosome.

- The position of the gene on the p or q arm. The position of a gene is based on a distinctive pattern of light and dark bands that appear when the chromosome is stained in a certain way. The position is usually designated by two digits (representing a region and a band), which are sometimes followed by a decimal point and one or more additional digits (representing sub-bands within a light or dark area). The number indicating the gene position increases with distance from the centromere. For example: 14q21 represents position 21 on the long arm of chromosome 14. 14q21 is closer to the centromere than 14q22.

Sometimes, the abbreviations "cen" or "ter" are also used to describe a gene's cytogenetic location. "Cen" indicates that the gene is very close to the centromere. For example, 16pcen refers to the short arm of chromosome 16 near the centromere. "Ter" stands for terminus, which indicates that the gene is very close to the end of the p or q arm. For example, 14qter refers to the tip of the long arm of chromosome 14. ("Tel" is also sometimes used to describe a gene's location. "Tel" stands for telomeres, which are at the ends of each chromosome. The abbreviations "tel" and "ter" refer to the same location.)

Molecular Location

The Human Genome Project, an international research effort completed in 2003, determined the sequence of base pairs for each human chromosome. This sequence information allows researchers to provide a more specific address than the cytogenetic location for many genes. A gene's molecular address pinpoints the location of that gene in terms of base pairs. It describes the gene's precise position on a chromosome and indicates the size of the gene. Knowing the molecular location also allows researchers to determine exactly how far a gene is from other genes on the same chromosome.

Different groups of researchers often present slightly different values for a gene's molecular location. Researchers interpret the sequence of the human genome using a variety of methods, which can result in small differences in a gene's molecular address. Genetics Home Reference presents data from NCBI for the molecular location of genes.

What are Gene Families?

A gene family is a group of genes that share important characteristics. In many cases, genes in a family share a similar sequence of DNA building blocks (nucleotides). These genes provide instructions for making products (such as proteins) that have a similar structure or function. In other cases, dissimilar genes are grouped together in a family because proteins produced from these genes work together as a unit or participate in the same process.

Classifying individual genes into families helps researchers describe how genes are related to each other. Researchers can use gene families to predict the function of newly identified genes based on their similarity to known genes. Similarities among genes in a family can also be used to predict where and when a specific gene is active (expressed). Additionally, gene families may provide clues for identifying genes that are involved in particular diseases. Sometimes not enough is known about a gene to assign it to an established family. In other cases, genes may fit into more than one family. No formal guidelines define the criteria for grouping genes together. Classification systems for genes continue to evolve as scientists learn more about the structure and function of genes and the relationships between them. Genetics Home Reference provides information about gene families including a brief description of each gene family and a list of the genes included in the family.

The HUGO Gene Nomenclature Committee (HGNC) has classified many human genes into families. Each grouping is given a name and symbol, and contains a table of the genes in that family. The textbook Human Molecular Genetics (second edition, 1999) provides background information on human gene families. The Gene Ontology database lists the protein products of genes by their location within the cell (cellular component), biological process, and molecular function. The Reactome database classifies the protein products of genes based on their participation in specific biological pathways. For example, this resource provides tables of genes involved in controlled cell death (apoptosis), cell division, and DNA repair.

Understanding Gene Testing

Genes- the chemical messages of heredity- constitute a blueprint of our possibilities and limitations. The legacy of generations of ancestors, our genes carry the key to our similarities and our uniqueness. When genes are working properly, our bodies develop and function smoothly. But should a single gene-even a tiny segment of a single gene- go awry, the effect can be dramatic: deformities and disease, even death. In the past 20 years, amazing new techniques have allowed scientists to learn a great deal about how genes work and how genes are linked to disease. Increasingly, researchers are able to identify mutations, changes within genes that can lead to specific disorders. Tests for gene mutations make it possible not only to detect diseases already in progress but also, in certain situations, to foresee diseases yet to come.

This new ability raises both high hopes and grave concerns. On the one hand, predictive gene testing holds out the possibility of saving thousands of lives through prevention or early detection. On the other, the implications of test results are enormous, not only for the individual but also for relatives who share this genetic legacy, and for society as a whole. This booklet presents key concepts and issues relevant to gene testing and answers questions that are frequently asked about the techniques, potential risks, and possible benefits of attempts to link genetic markers with disease. Words that appear underlined in the pages that follow are defined in the Glossary. Clicking on these words will take you to the glossary. From there press the browser's back button to return.

What are Genes?

Genes are working subunits of DNA. DNA is a vast chemical information database that carries the complete set of instructions for making all the proteins a cell will ever need. Each gene contains a particular set of instructions, usually coding for a particular protein.

DNA exists as two long, paired strands spiraled into the famous double helix. Each strand is made up of millions of chemical building blocks called bases. While there are only four different chemical bases in DNA (adenine, thymine, cytosine, and guanine), the order in which the bases occur determines the information available, much as specific letters of the alphabet combine to form words and sentences.

DNA resides in the core, or nucleus, of each of the body's trillions of cells. Every human cell (with the exception of mature red blood cells, which have no nucleus) contains the same DNA. Each cell has 46 molecules of double-stranded DNA. Each molecule is made up of 50 to 250 million bases housed in a chromosome. The DNA in each chromosome constitutes many genes (as well as vast stretches of noncoding DNA, the function of which is unknown). A gene is any given segment along the DNA that encodes instructions that allow a cell to produce a specific product- typically, a protein such as an enzyme- that initiates one specific action. There are between 50,000 and 100,000 genes, and every gene is made up of thousands, even hundreds of thousands, of chemical bases.

How do Genes Work?

Although each cell contains a full complement of DNA, cells use genes selectively. Some genes enable cells to make proteins needed for basic functions; dubbed housekeeping genes, they are active in many types of cells. Other genes, however, are inactive most of the time. Some genes play a role in early development of the embryo and are then shut down forever. Many genes encode proteins that are unique to a particular kind of cell and that give the cell its character- making a brain cell, say, different from a bone cell. A normal cell activates just the genes it needs at the moment and actively suppresses the rest.

RNA and DNA Revealed: New Roles, New Rules

For many years, when scientists thought about heredity, DNA was the first thing to come to mind. It's true that DNA is the basic ingredient of our genes and, as such, it often steals the limelight from RNA, the other form of genetic material inside our cells.

But, while they are both types of genetic material, RNA and DNA are rather different. The chemical units of RNA are like those of DNA, except that RNA has the nucleotide uracil (U) instead of thymine (T). Unlike double-stranded DNA, RNA usually comes as only a single strand. And the nucleotides in RNA contain ribose sugar molecules in place of deoxyribose. RNA is quite flexible—unlike DNA, which is a rigid, spiral-staircase molecule that is very stable. RNA can twist itself into a variety of complicated, three-dimensional shapes. RNA is also unstable in that cells constantly break it down and must continually make it fresh, while DNA is not broken down often. RNA's instability lets cells change their patterns of protein synthesis very quickly in response to what's going on around them.

Many textbooks still portray RNA as a passive molecule, simply a "middle step" in the cell's gene-reading activities. But that view is no longer accurate. Each year, researchers unlock new secrets about RNA. These discoveries reveal that it is truly a remarkable molecule and a multi-talented actor in heredity.

Today, many scientists believe that RNA evolved on the Earth long before DNA did. Researchers hypothesize—obviously, no one was around to write this down—that RNA was a major participant in the chemical reactions that ultimately spawned the first signs of life on the planet.

RNA World

At least two basic requirements exist for making a cell: the ability to hook molecules together and break them apart, and the ability to replicate, or copy itself, from existing information. RNA probably helped to form the first cell. The first organic molecules, meaning molecules containing carbon, most likely arose out of random collisions of gases in the Earth's primitive atmosphere, energy from the Sun, and heat from naturally occurring radioactivity. Some scientists think that in this primitive world, RNA was a critical molecule because of its ability to lead a double life: to store information and to conduct chemical reactions. In other words, in this world, RNA served the functions of both DNA and proteins.

What does any of this have to do with human health? Plenty, it turns out. Today's researchers are harnessing some of RNA's flexibility and power. For example, through a strategy he calls directed evolution, molecular engineer Ronald R. Breaker of Yale University is developing ways to create entirely new forms of RNA and DNA that both work as enzymes. Recently, Breaker and others have also uncovered a hidden world of RNAs that play a major role in controlling gene activity, a job once thought to be performed exclusively by proteins. These RNAs, which the scientists named riboswitches, are found in a wide variety of bacteria and other organisms. This discovery has led Breaker to speculate that new kinds of antibiotic medicines could be developed to target bacterial riboswitches.

Molecular Editor

Scientists are learning of another way to customize proteins: by RNA editing. Although DNA sequences spell out instructions for producing RNA and proteins, these instructions aren't always followed precisely. Editing a gene's mRNA, even by a single chemical letter, can radically change the resulting protein's function. Nature likely evolved the RNA editing function as a way to get more proteins out of the same number of genes. For example, researchers have found that the mRNAs for certain proteins important for the proper functioning of the nervous system are particularly prone to editing.

It may be that RNA editing gives certain brain cells the capacity to react quickly to a changing environment. Which molecules serve as the editor and how does this happen? Brenda Bass of the University of Utah School of Medicine in Salt Lake City studies one particular class of editors called adenosine deaminases. These enzymes "retype" RNA letters at various places within an mRNA transcript.

They do their job by searching for characteristic RNA shapes. Telltale twists and bends in folded RNA molecules signal these enzymes to change the RNA sequence, which in turn changes the protein that gets made. Bass' experiments show that RNA editing occurs in a variety of organisms, including people. Another interesting aspect of editing is that certain disease-causing microorganisms, such as some forms of parasites, use RNA editing to gain a survival edge when living in a human host. Understanding the details of this process is an important area of medical research.

Small But Powerful

Recently, molecules called microRNAs have been found in organisms as diverse as plants, worms, and people. The molecules are truly "micro," consisting of only a few dozen nucleotides, compared to typical human mRNAs that are a few thousand nucleotides long. What's particularly interesting about microRNAs is that many of them arise from DNA that used to be considered merely filler material.

How do these small but important RNA molecules do their work? They start out much bigger but get trimmed by cellular enzymes, including one aptly named Dicer. Like tiny pieces of Velcro®, microRNAs stick to certain mRNA molecules and stop them from passing on their protein-making instructions. First discovered in a roundworm model system some microRNAs help determine the organism's body plan. In their absence, very bad things can happen. For example, worms engineered to lack a microRNA called let-7 develop so abnormally that they often rupture and practically break in half as the worm grows.

Perhaps it is not surprising that since microRNAs help specify the timing of an organism's developmental plan, the appearance of the microRNAs themselves is carefully timed inside a developing organism. Biologists, including Amy Pasquinelli of the University of California, San Diego, are currently figuring out how microRNAs are made and cut to size, as well as how they are produced at the proper time during development.

MicroRNA molecules also have been linked to cancer. For example, Gregory Hannon of the Cold Spring Harbor Laboratory on Long Island, New York, found that certain microRNAs are associated with the severity of the blood cancer B-cell lymphoma in mice. Since the discovery of microRNAs in the first years of the 21st century, scientists have identified hundreds of them that likely exist as part of a large family with similar nucleotide sequences. New roles for these molecules are still being found.

RNA Interference (RNAi)

Healthy Interference

RNA controls genes in a way that was only discovered recently: a process called RNA interference, or RNAi. Although scientists identified RNAi less than 10 years ago, they now know that organisms have been using this trick for millions of years. Researchers believe that RNAi arose as a way to reduce the production of a gene's encoded protein for purposes of fine-tuning growth or self-defence. When viruses infect cells, for example, they command their host to produce specialized RNAs that allow the virus to survive and make copies of itself. Researchers believe that RNAi eliminates unwanted viral RNA, and some speculate that it may even play a role in human immunity.

Oddly enough, scientists discovered RNAi from a failed experiment! Researchers investigating genes involved in plant growth noticed something strange: When they tried to turn petunia flowers purple by adding an extra "purple" gene, the flowers bloomed white instead. This result fascinated researchers, who could not understand how adding genetic material could somehow get rid of an inherited trait. The mystery remained unsolved until, a few years later, two geneticists studying development saw a similar thing happening in lab animals.

The researchers, Andrew Z. Fire, then of the Carnegie Institution of Washington in Baltimore and now at Stanford University, and Craig Mello of the University of Massachusetts Medical School in Worcester, were trying to block the expression of genes that affect cell growth and tissue formation in roundworms, using a molecular tool called antisense RNA. To their surprise, Mello and Fire found that their antisense RNA tool wasn't doing much at all. Rather, they determined, a doublestranded contaminant produced during the synthesis of the single-stranded antisense RNA interfered with gene expression. Mello and Fire named the process RNAi, and in 2006 were awarded the Nobel Prize in physiology or medicine for their discovery.

Further experiments revealed that the doublestranded RNA gets chopped up inside the cell into much smaller pieces that stick to mRNA and block its action, much like the microRNA pieces of Velcro discussed above. Today, scientists are taking a cue from nature and using RNAi to explore biology. They have learned, for example, that the process is not limited to worms and plants, but operates in humans too. Medical researchers are currently testing new types of RNAi-based drugs for treating conditions such as macular degeneration, the leading cause of blindness, and various infections, including those caused by HIV and herpes virus.

Dynamic DNA

A good part of who we are is "written in our genes," inherited from Mom and Dad. Many traits, like red or brown hair, body shape, and even some personality quirks, are passed on from parent to offspring. But genes are not the whole story. Where we live, how much we exercise, what we eat: These and many other environmental factors can all affect how our genes get expressed. You know that changes in DNA and RNA can produce changes in proteins. But additional control happens at the level of DNA, even though these changes do not alter DNA directly. Inherited factors that do not change the DNA sequence of nucleotides are called epigenetic changes, and they too help make each of us unique.

Epigenetic means, literally, "upon" or "over" genetics. It describes a type of chemical reaction that can alter the physical properties of DNA without changing its sequence. These changes make genes either more or less likely to be expressed. Currently, scientists are following an intriguing course of discovery to identify epigenetic factors that, along with diet and other environmental influences, affect who we are and what type of illnesses we might get.

Secret Code

DNA is spooled up compactly inside cells in an arrangement called chromatin. This packaging is critical for DNA to do its work. Chromatin consists of long strings of DNA spooled around a compact assembly of proteins called histones. One of the key functions of chromatin is to control access to genes, since not all genes are turned on at the same time. Improper expression of growth-promoting genes, for example, can lead to cancer, birth defects, or other health concerns. Many years after the structure of DNA was determined, researchers used a powerful device known as an electron microscope to take pictures of chromatin fibres. Upon viewing chromatin up close, the researchers described it as "beads on a string," an image still used today. The beads were the histone balls, and the string was DNA wrapped around the histones and connecting one bead to the next. Decades of study eventually revealed that histones have special chemical tags that act like switches to control access to the DNA. Flipping these switches, called epigenetic markings, unwinds the spooled DNA so the genes can be transcribed.

The observation that a cell's gene-reading machinery tracks epigenetic markings led C. David Allis, who was then at the University of Virginia Health Sciences Centre in Charlottesville and now works at the Rockefeller University in New York City, to coin a new phrase, the "histone code." He and others believe that the histone code plays a major role in determining which proteins get made in a cell. Flaws in the histone code have been associated with several types of cancer, and researchers are actively pursuing the development of medicines to correct such errors.

Genetics and You: The Genetics of Anticipation

Occasionally, unusual factors influence whether or not a child will be born with a genetic disease. An example is the molecular error that causes Fragile X syndrome, a rare condition associated with mental retardation. The mutation leading to a fragile X chromosome is not a typical DNA typing mistake, in which nucleotides are switched around or dropped, or one of them is switched for another nucleotide. Instead, it is a kind of stutter by the DNA polymerase enzyme that copies DNA. This stutter creates a string of repeats of a DNA sequence that is composed of just three nucleotides, CGG. The number of triplet repeats seems to increase as the chromosome is passed down through several generations. Thus, the grandsons of a man with a fragile X chromosome, who is not himself affected, have a 40 percent risk of retardation if they inherit the repeat-containing chromosome. The risk for great-grandsons is even higher: 50 percent.

Intrigued by the evidence that triplet repeats can cause genetic disease, scientists have searched for other examples of disorders associated with the DNA expansions. To date, more than a dozen such disorders have been found, and all of them affect the nervous system. Analysis of the rare families in which such diseases are common has revealed that expansion of the triplet repeats is linked to something called genetic anticipation, when a disease's symptoms appear earlier and more severely in each successive generation.

Battle of the Sexes

A process called imprinting, which occurs naturally in our cells, provides another example of how epigenetics affects gene activity. With most genes, the two copies work exactly the same way. For some mammalian genes, however, only the mother's or the father's copy is switched on regardless of the child's gender. This is because the genes are chemically marked, or imprinted, during the process that generates eggs and sperm. As a result, the embryo that emerges from the joining of egg and sperm can tell whether a gene copy came from Mom or Dad, so it knows which copy of the gene to shut off.

One example of an imprinted gene is insulin-like growth factor 2 (Igf2), a gene that helps a mammalian fetus grow. In this case, only the father's copy of Igf2 is expressed, and the mother's copy remains silent (is not expressed) throughout the life of the offspring. Scientists have discovered that this selective silencing of Igf2 and many other imprinted genes occurs in all placental mammals (all except the platypus, echidna, and marsupials) examined so far, but not in birds.

Why would nature tolerate a process that puts an organism at risk because only one of two copies of a gene is working? The likely reason, many researchers believe, is that mothers and fathers have competing interests, and the battle- field is DNA! The scenario goes like this: It is in a father's interest for his embryos to get bigger faster, because that will improve his offspring's chances of survival after birth. The better an individual's chance of surviving infancy, the better its chance of becoming an adult, mating, and passing its genes on to the next generation. Of course mothers want strong babies, but unlike fathers, mothers provide physical resources to embryos during pregnancy. Over her lifetime, a female is likely to be pregnant several times, so she needs to divide her resources among a number of embryos in different pregnancies.

Researchers have discovered over 200 imprinted genes in mammals since the first one was identified in 1991. We now know that imprinting controls some of the genes that have an important role in regulating embryonic and fetal growth and allocating maternal resources. Not surprisingly, mutations in these genes cause serious growth disorders. Marisa Bartolomei of the University of Pennsylvania School of Medicine in Philadelphia is trying to figure out how Igf2 and other genes become imprinted and stay silent throughout the life of an individual. She has already identified sequences within genes that are essential for imprinting. Bartolomei and other researchers have shown that these sequences, called insulators, serve as "landing sites" for a protein that keeps the imprinted gene from being transcribed.

Starting at the End

When we think of DNA, we think of genes. However, some DNA sequences are different: They don't encode RNAs or proteins. Introns. Another example is telomeres—the ends of chromosomes. There are no genes in telomeres, but they serve an essential function. Like shoelaces without their tips, chromosomes without telomeres unravel and fray. And without telomeres, chromosomes stick to each other and cause cells to undergo harmful changes like dividing abnormally.

Researchers know a good deal about telomeres, dating back to experiments performed in the 1970s by Elizabeth Blackburn, a basic researcher who was curious about some of the fundamental events that take place within cells. At the time, Blackburn, now at the University of California, San Francisco, was working with Joseph Gall at Yale University. For her experimental system, she chose a single-celled, pond-dwelling organism named *Tetrahymena*. These tiny, pear-shaped creatures are covered with hairlike cilia that they use to propel themselves through the water as they devour bacteria and fungi. *Tetrahymena* was a good organism for Blackburn's experiments because it has a large number of chromosomes—which means it has a lot of telomeres!

Her research was also perfectly timed, because methods for sequencing DNA were just being developed. Blackburn found that *Tetrahymena's* telomeres had an unusual nucleotide sequence: TTGGGG, repeated about 50 times per telomere. Since then, scientists have discovered that the telomeres of almost all organisms have repeated sequences of DNA with lots of Ts and Gs. In human and mouse telomeres, for example, the repeated sequence is TTAGGG.

The number of telomere repeats varies enormously, not just from organism to organism but in different cells of the same organism and even within a single cell over time. Blackburn reasoned that the repeat number might vary if cells had an enzyme that added copies of the repeated sequence to the telomeres of some but not all chromosomes. With her then-graduate student Carol Greider, now at Johns Hopkins University, Blackburn hunted for the enzyme. The team found it and Greider named it telomerase. The telomerase enzyme, it turns out, consisted of a protein and an RNA component, which the enzyme uses as a template for copying the repeated DNA sequence. What is the natural function of telomerase? As cells divide again and again, their telomeres get shorter.Most normal cells stop dividing when their telomeres wear down to a certain point, and eventually the cells die. Telomerase can counteract the shortening. By adding DNA to telomeres, telomerase rebuilds the telomere and resets the cell's molecular clock.

The discovery of telomerase triggered new ideas and literally thousands of new studies. Many researchers thought that the enzyme might play important roles in cancer and aging. Researchers were hoping to find ways to turn telomerase on so that cells would continue to divide (to grow extra cells for burn patients, for example), or *off* so that cells would stop dividing (to stop cancer, for instance). So far, they have been unsuccessful. Although it is clear that telomerase and cellular aging are related, researchers do not know whether telomerase plays a role in the normal cellular aging process or in diseases like cancer. Recently, however, Blackburn and a team of other scientists discovered that chronic stress and the perception that life is stressful affect telomere length and telomerase activity in the cells of healthy women. Blackburn and her coworkers are currently conducting a long-term, follow-up study to confirm these intriguing results.

The other Human Genome

Before you think everything's been said about DNA, there's one little thing we didn't mention: Some of the DNA in every cell is quite different from the DNA that we've been talking about up to this point. This special DNA isn't in chromosomes— it isn't even inside the cell's nucleus where all the chromosomes are! So where is this special DNA? It's inside mitochondria, the organelles in our cells that produce the energy-rich molecule adenosine triphosphate, or ATP. Mendel knew nothing of mitochondria, since they weren't discovered until late in the 19th century. And it wasn't until the 1960s that researchers discovered the mitochondrial genome, which is circular like the genomes of bacteria. In human cells, mitochondrial DNA makes up less than 1 percent of the total DNA in each of our cells. The mitochondrial genome is very small—containing only about three dozen genes. These encode a few of the proteins that are in the mitochondrion, plus a set of ribosomal RNAs used for synthesizing proteins for the organelle. Mitochondria need many more proteins though, and most of these are encoded by genes in the nucleus. Thus, the energy-producing capabilities of human mitochondria—a vital part of any cell's everyday health—depend on coordinated teamwork among hundreds of genes in two cellular neighborhoods: the nucleus and the mitochondrion. Mitochondrial DNA gets transcribed and the RNA is translated by enzymes that are very different from those that perform this job for genes in our chromosomes. Mitochondrial enzymes look and act much more like those from bacteria, which is not surprising because mitochondria are thought to have descended from free-living bacteria that were engulfed by another cell over a billion years ago.

Scientists have linked mitochondrial DNA defects with a wide range of age-related diseases including neurodegenerative disorders, some forms of heart disease, diabetes, and various cancers. It is still unclear, though, whether damaged mitochondria are a symptom or a cause of these health conditions. Scientists have studied mitochondrial DNA for another reason: to understand the history of the human race. Unlike our chromosomal DNA, which we inherit from both parents, we get all of our mitochondrial DNA from our mothers.

Thus, it is possible to deduce who our maternal ancestors were by tracking the inheritance of mutations in mitochondrial DNA. For reasons that are still not well understood, mutations accumulate in mitochondrial DNA very slowly compared to chromosomal DNA. So, it's possible to trace your maternal ancestry way back beyond any relatives you may know by name—all the way back to "African Eve," the ancestor of us all!

The Tools of Genetics: Recombinant DNA and Cloning

To splice a human gene (in this case, the one for insulin) into a plasmid, scientists take the plasmid out of an *E. coli* bacterium, cut the plasmid with a restriction enzyme, and splice in insulin-making human DNA. The resulting hybrid plasmid can be inserted into another *E. coli* bacterium, where it multiplies along with the bacterium. There, it can produce large quantities of insulin. In the early 1970s, scientists discovered that they could change an organism's genetic traits by putting genetic material from another organism into its cells. This discovery, which caused quite a stir, paved the way for many extraordinary accomplishments in medical research that have occurred over the past 35 years.

How do scientists move genes from one organism to another? The cutting and pasting gets done with chemical scissors: enzymes, to be specific. Take insulin, for example. Let's say a scientist wants to make large quantities of this protein to treat diabetes. She decides to transfer the human gene for insulin into a bacterium, *Escherichia coli*, or *E. coli*, which is commonly used for genetic research. That's because E. coli reproduces really fast, so after one bacterium gets the human insulin gene, it doesn't take much time to grow millions of bacteria that contain the gene. The first step is to cut the insulin gene out of a copied, or "cloned," version of the human DNA using a special bacterial enzyme from bacteria called a restriction endonuclease. (The normal role of these enzymes in bacteria is to chew up the DNA of viruses and other invaders.) Each restriction enzyme recognizes and cuts at a different nucleotide sequence, so it's possible to be very precise about DNA cutting by selecting one of several hundred of these enzymes that cuts at the desired sequence. Most restriction endonucleases make slightly staggered incisions, resulting in "sticky ends," out of which one strand protrudes.

The next step in this example is to splice, or paste, the human insulin gene into a circle of bacterial DNA called a plasmid. Attaching the cut ends together is done with a different enzyme (obtained from a virus), called DNA ligase. The sticky ends join back together kind of like jigsaw puzzle pieces. The result: a cut-and-pasted mixture of human and bacterial DNA.

The last step is putting the new, recombinant DNA back into *E. coli* and letting the bacteria reproduce in a petri dish. Now, the scientist has a great tool: a version of *E. coli* that produces lots of human insulin that can be used for treating people with diabetes.

So, what is cloning? Strictly speaking, it's making many copies of a gene—in the example above, *E. coli* is doing the cloning. However, the term cloning is more generally used to refer to the entire process of isolating and manipulating a gene. Dolly the cloned sheep contained the identical genetic material of another sheep. Thus, researchers refer to Dolly as a clone.

Rules Governing RNA's Anatomy Revealed

"RNA is a very floppy molecule that often functions by binding to something else and then radically changing shape," said Al-Hashimi, who is the Robert L. Kuczkowski Professor of Chemistry and a professor of biophysics. These shape changes, in turn, trigger other processes or cascades of events, such as turning specific genes on or off. Because of the RNA molecule's mercurial nature, "you can't really define it as having a single structure," Al-Hashimi said. "It has many possible orientations, and different orientations are stabilized under different conditions, such as the presence of particular drug molecules."

A major goal in structural biology and biophysics is to be able to predict not only the complex three-dimensional shapes that RNA assumes (which are dictated by the order of its nucleic acid building blocks), but also the various shapes RNA takes on after binding to other molecules such as proteins and small-molecule drugs. Further, researchers would like to be able to manipulate the 3-D structure and resulting activity of RNA by tweaking the drug molecules with which it interacts. But to do that, they need to understand the rules that govern the anatomy of RNA. The quest has parallels to the study of human anatomy, Al-Hashimi said. "Your body has a specific shape that changes predictably when you are walking or when you are catching a ball; we want to be able to understand these anatomical rules in RNA."

Manipulating RNA is a much sought-after goal, given the recent explosion in vital cellular roles ascribed to RNA and the growing number of diseases that are linked to RNA malfunction. RNA performs many of its roles by serving as a switch that changes shape in response to cellular signals, prompting appropriate reactions in response. The versatile molecule also is essential to retroviruses such as HIV, which have no DNA and instead rely on RNA to both transport and execute genetic instructions for everything the virus needs to invade and hijack its host.

In earlier work, Al-Hashimi's team determined that rather than changing shape in response to encounters with drug molecules, RNA goes through a predictable course of shape changes on its own. Drug molecules simply "wait for" the right shape and attach to RNA when the RNA assumes the particular drug's preferred orientation, Al Hashimi said. But what rules control the predictable path of shapes the RNA molecule assumes? And are those rules the same for all sorts of RNA molecules? In the current work, Al-Hashimi's team investigated those questions.

"RNA is very similar to the human body in its construction, in that it's made up of limbs that are connected at joints," Al-Hashimi said. The limbs are the familiar, ladder-like double helix structures, and the joints are flexible junctions. The prevailing view was that interactions among loopy structures at the tips of the limbs played a role in defining the molecule's overall 3-D shape, much as a handshake defines the orientation of two arms, but Al-Hashimi's group decided to look at things from a different perspective. "We wondered if the junctions themselves might provide the definition," Al-Hashimi said.

"If you look at your arm, you'll notice that you can't move it, relative to your shoulder, in just any way; it's confined to a certain pathway because of the joint's geometry. We wondered if the same thing might be true of RNA."

To investigate that possibility, the researchers turned to a database of RNA structures and found that all structures with two helices linked by a particular type of junction called a trinucleotide bulge fell along the same pathway.

The team then went on to explore structures of RNA molecules with other kinds of junctions. All were confined to similar pathways, but the precise pathway of a given RNA depended upon structural features of its junction. Just as anatomical features of our shoulders, elbows, hips and knees define the range of motion of our arms and legs, the anatomy of RNA's junctions dictates the motion of its helices.

Next, Al-Hashimi and coworkers wanted to understand how drug molecules cause RNA molecules to freeze in specific positions. In earlier work with an RNA molecule known as TAR, which is critical for replication of HIV and thus a key target for anti-HIV drugs, the researchers had found that certain drug molecules froze the RNA molecule in a nearly straight position, while others trapped the molecule in a bent conformation and still others captured positions between the two extremes. But because that project involved a wide variety of drug molecules, it was hard to figure out why certain ones preferred certain orientations.

To explore the issue more methodically, Al-Hashimi's group used a series of aminoglycosides (antibiotics that are known to target RNA) that systematically differed from one another in charge, size and other chemical properties. Size turned out to be the key: bigger aminoglycosides froze RNA in more bent positions; smaller ones favoured straighter RNA structures. Looking more closely, the researchers discovered that the aminoglycoside molecule nestles between two helices and acts like wedge, forcing the helices apart. Examination of other RNA structures bound to small molecules revealed that this rule is not specific to TAR but a more general feature of RNA-small molecule interactions.

"With these findings, it now should be possible to predict gross features of RNA 3-D shapes based only on their secondary structure, which is far easier to determine than is 3-D structure," Al-Hashimi said. "This will make it possible to gain insights into the 3-D shapes of RNA structures that are too large or complicated to be visualized by experimental techniques such as X-ray crystallography and NMR spectroscopy. The anatomical rules also provide a blueprint for rationally manipulating the structure and thus the activity of RNA, using small molecules in drug design efforts and also for engineering RNA sensors that change structure in user-prescribed ways."

Discovery of the Hybrid Helix and the First DNA-RNA Hybridization

Here we describe early research on RNA structure and the discovery of the DNA-RNA hybrid helix, a key component of information transfer. More than 50 years ago it was realized that the interaction between DNA and RNA was at the core of molecular biology. The problem was chemical in nature: could two different types of molecules interact and serve in the transmission of biological information?

In the 1950s it was widely assumed that "DNA makes RNA, RNA makes protein." This was not based on experimental evidence that DNA and RNA could combine but was more in the nature of an intuitive belief. However, by early 1960 I was finally able to carry out a direct experiment, the first DNA-RNA hybridization. In 1960, messenger RNA was still 1 year in the future, and there was not a great deal of understanding of the major components of information transfer. The DNA double helix proposed by Watson and Crick in 1953 clearly suggested that information was contained in the order of nucleotides, but during the 1950s our understanding of RNA was fragmentary. The origins of the 1960 experiment go back to the mid-1950s. In 1954, while at Caltech, Jim Watson and I had been trying to find out if RNA by itself could form a double helix, but the fibre diffraction studies of RNA gave inconclusive results. The fuzzy diffraction patterns all looked alike, but unlike DNA, the RNA base ratios were all different in the samples we examined. It was clear that RNA was more complex than DNA, and its structure was unknown. In their 1953 paper Watson and Crick pointed out that it was probably impossible to form their double helix with ribose due to van der Waals interference of the 22-OH with the structure. Thus, it was likely to be different.

4

Plant Reproduction

If you are an organism, you will need to reproduce. Otherwise, there will be no more of your species and the species will die off. You may have heard of endangered animals. There are also endangered plants. These endangered species have very few individuals left and scientists/naturalists are working together to make sure the species don't become extinct. We talked a little about reproduction when we discussed meiosis in the cells tutorials. Reproduction is one of two things.

(1) One cell can split into two, giving you two identical cells. That type is asexual reproduction.

(2) The second type is when two cells, each with half of the DNA needed, combine and create a living cell. That type is sexual reproduction.

When plants hit a point in evolution, the second is the one that occurs more often.

Making more Mosses

Sporophytes are the reproductive structures you will find in mosses. They are actually a phase of the moss life cycle that feeds off the green parent plant (the gametophyte). The sporophyte is a stalk that grows after the haploid sperm of one moss plant is able to mix with the haploid egg of a female moss plant. The resulting diploid cell grows into the sporophyte stalk. When ready, spores stored in the sporophyte are released and they grow into new moss plants.

Confiers and their Cones

While there are male and female mosses, conifers produce two types of cones on the same tree. One of the cone types gives off pollen (the staminate cone). The other type of cone catches the pollen if the wind is moving in the right direction. Better yet, the wind blows the pollen to another conifer of the same species, and a cone (called the ovulate cone) catches the pollen. Again, the pollen and megaspore (receiving haploid cell) are haploid and combine to form a diploid cell. That diploid cell grows into a zygote (baby conifer) that eventually lives in a seed.

Flowers and Pollen

The most advanced of the plants have their own way of sexually reproducing. It is a very fancy and very complex process. Plants that rely on flowers for reproduction are also very dependent on outside help such as insects and animals. While conifers have the two structures on one tree, flowering plants went one step further and put the devices that make and receive pollen in the same structure. How does that help? A bee might go to one flower and get a little pollen on its back. If it goes to another flower of the same species, that pollen can land on the stigma. From that point, one haploid male nucleus combines with a female nucleus and the other haploid male nucleus combines with a polar nucleus. If successful, an embryo and seed/fruit develop respectively.

Reproduction in Plants

Plant reproduction is the process by which plants generate new individuals, or offspring. Reproduction is either sexual or asexual. Sexual reproduction is the formation of offspring by the fusion of gametes. Asexual reproduction is the formation of offspring without the fusion of gametes. Sexual reproduction results in offspring genetically different from the parents. Asexual offspring are genetically identical except for mutation. In higher plants, offspring are packaged in a protective seed, which can be long lived and can disperse the offspring some distance from the parents. In flowering plants (angiosperms), the seed itself is contained inside a fruit, which may protect the developing seeds and aid in their dispersal.

Sexual Reproduction in Angiosperms: Ovule Formation

All plants have a life cycle that consists of two distinct forms that differ in size and the number of chromosomes per cell. In flowering plants, the large, familiar form that consists of roots, shoots, leaves, and reproductive structures (flowers and fruit) is diploid and is called the sporophyte. The sporophyte produces haploid microscopic gametophytes that are dependent on tissues produced by the flower. The reproductive cycle of a flowering plant is the regular, usually seasonal, cycling back and forth from sporophyte to gametophyte.

The flower produces two kinds of gametophytes, male and female. The female gametophyte arises from a cell within the ovule, a small structure within the ovary of the flower. The ovary is a larger structure within the flower that contains and protects usually many ovules. Flowering plants are unique in that their ovules are entirely enclosed in the ovary. The ovary itself is part of a larger structure called the carpel, which consists of the stigma, style, and ovary. Each ovule is attached to ovary tissue by a stalk called the funicle. The point of attachment of the funicle to the ovary is called the placenta.

As the flower develops from a bud, a cell within an ovule called the archespore enlarges to form an embryo-sac mother cell (EMC). The EMC divides by meiosis to produce four megaspores. In this process the number of chromosomes is reduced from two sets in the EMC to one set in the megaspores, making the megaspores haploid. Three of the four megaspores degenerate and disappear, while the fourth divides mitotically three times to produce eight haploid cells. These cells together constitute the female gametophyte, called the embryo sac.

The eight embryo sac cells differentiate into two synergids, three antipodal cells, two fused endosperm nuclei, and an egg cell. The mature embryo sac is situated at the outer opening (micropyle) of the ovule, ready to receive the sperm cells delivered by the male gametophyte.

Pollen

The male gametophyte is the mature pollen grain. Pollen is produced in the anthers, which are attached at the distal end of filaments. The filament and anther together constitute the stamen, the male sex organ. Flowers usually produce many stamens just inside of the petals. As the flower matures, cells in the anther divide mitotically to produce pollen mother cells (PMC). The PMCs divide by meiosis to produce haploid microspores in groups of four called tetrads. The microspores are housed within a single layer of cells called the tapetum, which provides nutrition to the developing pollen grains.

Each microspore develops a hard, opaque outer layer called the exine, which is constructed from a lipoprotein called sporopollenin. The exine has characteristic pores, ridges, or projections that can often be used to identify a species, even in fossil pollen. The microspore divides mitotically once or twice to produce two or three haploid nuclei inside the mature pollen grain. Two of the nuclei function as sperm nuclei that can eventually fuse with the egg and endosperm nuclei of the embryo sac, producing an embryo and endosperm, respectively.

For sexual fusion to take place, however, the pollen grain must be transported to the stigma, which is a receptive platform on the top of the style, an elongated extension on top of the carpel(s). Here the moist surface or chemicals cause the pollen grain to germinate. Germination is the growth of a tube from the surface of a pollen grain. The tube is a sheath of pectin, inside of which is a solution of water, solutes, and the two or three nuclei, which lack any cell walls. Proper growth of the pollen tube requires an aqueous solution of appropriate solute concentration, as well as nutrients such as boron, which may aid in its synthesis of pectin.

At the apex of the tube are active ribosomes and endoplasmic reticulum (types of cell organelles) involved in protein synthesis. Pectinase and a glucanase (both enzymes that break down carbohydrates) probably maintain flexibility of the growing tube and aid in penetration. The pollen tube apex also releases ribonucleic acid (RNA) and ribosomes into the tissues of the style. The tube grows to eventually reach the ovary, where it may travel along intercellular spaces until it reaches a placenta. Through chemical recognition, the pollen tube changes its direction of growth and penetrates through the placenta to the ovule. Here the tube reaches the embryo sac lying close to the micropyle, and sexual fertilization takes place.

Double Fertilization

Fertilization in flowering plants is unique among all known organisms, in that not one but two cells are fertilized, in a process called double fertilization. One sperm nucleus in the pollen tube fuses with the egg cell in the embryo sac, and the other sperm nucleus fuses with the diploid endosperm nucleus. The fertilized egg cell is a zygote that develops into the diploid embryo of the sporophyte. The fertilized endosperm nucleus develops into the triploid endosperm, a nutritive tissue that sustains the embryo and seedling. The only other known plant group exhibiting double fertilization is the Gnetales in the genus *Ephedra,* a nonflowering seed plant. However, in this case the second fertilization product degenerates and does not develop into endosperm. Double fertilization begins when the pollen tube grows into one of the two synergid cells in the embryo sac, possibly as a result of chemical attraction to calcium. After penetrating the synergid, the apex of the pollen tube breaks open, releasing the two sperm nuclei and other contents into the synergid. As the synergid degenerates, it envelops the egg and endosperm cells, holding the two sperm nuclei close and the other expelled contents of the pollen tube. The egg cell then opens and engulfs the sperm cell, whose membrane breaks apart and allows the nucleus to move near the egg nucleus. The nuclear envelopes then disintegrate, and the two nuclei combine to form the single diploid nucleus of the zygote. The other sperm cell fuses with the two endosperm nuclei, forming a single triploid cell, the primary endosperm cell, which divides mitotically into the endosperm tissue.

Double fertilization and the production of endosperm may have contributed to the great ecological success of flowering plants by accelerating the growth of seedlings and improving survival at this vulnerable stage. Faster seedling development may have given flowering plants the upper hand in competition with gymnosperm seedlings in some habitats, leading to the abundance of flowering plants in most temperate and tropical regions. Gymnosperms nevertheless are still dominant at higher elevations and latitudes, and at low elevations in the Pacific Northwest coniferous forests, such as the coastal redwoods. The reasons for these patterns are still controversial.

The Seed

The seed is the mature, fertilized ovule. After fertilization, the haploid cells of the embryo sac disintegrate. The maternally derived diploid cells of the ovule develop into the hard, water-resistant outer covering of the seed, called the testa, or seed coat. The diploid zygote develops into the embryo, and the triploid endosperm cells multiply and provide nutrition. The testa usually shows a scar called the hilum where the ovule was originally attached to the funicle. In some seeds a ridge along the testa called the raphe shows where the funicle originally was pressed against the ovule. The micropyle of the ovule usually survives as a small pore in the seed coat that allows passage of water during germination of the seed.

In some species, the funicle develops into a larger structure on the seed called an aril, which is often brightly colored, juicy, and contains sugars that are consumed by animals that may also disperse the seed (as in nutmeg, arrowroot, oxalis, and castor bean). This is distinct from the fruit, which forms from the ovary itself.

The embryo consists of the cotyledon(s), epicotyl, and hypocotyl. The cotyledons resemble small leaves, and are usually the first photosynthetic organs of the plant. The portion of the embryo above the cotyledons is the epicotyl, and the portion below is the hypocotyl. The epicotyl is an apical meristem that produces the shoot of the growing plant and the first true leaves after germination.

The hypocotyl develops into the root. Often the tip of the hypocotyl, the radicle, is the first indication of germination as it breaks out of the seed. Flowering plants are classified as monocotyledons or dicotyledons (most are now called eudicots) based on the number of cotyledons produced in the embryo. Common monocotyledons include grasses, sedges, lilies, irises, and orchids; common dicotyledons include sunflowers, roses, legumes, snapdragons, and all nonconiferous trees.

The endosperm may be consumed by the embryo, as in many legumes, which use the cotyledons as a food source during germination. In other species the endosperm persists until germination, when it is used as a food reserve. In grains such as corn and wheat, the outer layer of the endosperm consists of thick-walled cells called aleurone, which are high in protein.

The Fruit

The fruit of a flowering plant is the mature ovary. As seeds mature, the surrounding ovary wall forms a protective structure that may aid in dispersal. The surrounding ovary tissue is called the pericarp and consists of three layers. From the outside to inside, these layers are the exocarp, mesocarp, and endocarp. The exocarp is usually tough and skinlike. The mesocarp is often thick, succulent, and sweet. The endocarp, which surrounds the seeds, may be hard and stony, as in most species with fleshy fruit, such as apricots.

A fruit is termed simple if it is produced by a single ripened ovary in a single flower (apples, oranges, apricots). An aggregated fruit is a cluster of mature ovaries produced by a single flower (blackberries, raspberries, strawberries). A multiple fruit is a cluster of many ripened ovaries on separate flowers growing together in the same inflorescence (pineapple, mulberry. A simple fruit may be fleshy or dry. A fleshy simple fruit is classified as a berry (grape, tomato, papaya), pepo (cucumber, watermelon, pumpkin), hesperidium (orange), drupe (apricot), or pome (apple). Dry simple fruits have a dry pericarp at maturity. They may or may not dehisce, or split, along a seam to release the seeds. A dehiscent dry fruit is classified as legume or pod (pea, bean), silique or silicle (mustard), capsule (poppy, lily), or follicle (milkweed, larkspur, columbine). An indehiscent dry fruit that does not split to release seeds is classified as an achene (sunflower, buttercup, sycamore), grain or caryopsis (grasses such as corn, wheat, rice, barley), schizocarp (carrot, celery, fennel), winged samara (maple, ash, elm), nut (acorn, chestnut, hazelnut), or utricle (duckweed family). Some fruiting bodies contain non-ovary tissue and are sometimes called pseudocarps. The sweet flesh of apples and pears, for example, is composed not of the pericarp but the receptacle, or upper portion, of the flowering shoot to which petals and other floral organs are attached.

Fruiting bodies of all kinds function to protect and disperse the seeds they contain. Protection can be physical (hard coverings) or chemical (repellents of seed predators). Sweet, fleshy fruits are attractive food for birds and mammals that consume seeds along with the fruit and pass the seeds intact in their fecal matter, which can act as a fertilizer. Dry fruits are usually adapted for wind dispersal of seeds, as for example with the assistance of winglike structures or a fluffy pappus that provides buoyancy. The diversity of fruiting bodies reflects in part the diversity of dispersal agents in the environment, which select for different fruit size, shape, and chemistry.

Pollination and Pollinators

Pollination is the movement of pollen from the stamens to the stigma, where germination and growth of the pollen tube occur. Most (approximately 96 percent) of all flowering plant species are hermaphroditic (possess both sexual functions within a plant, usually within every flower), and thus an individual can be pollinated by its own pollen or by pollen from another individual. Seed produced through self-pollination ("selfed" seed) is often inferior in growth, survival, and fecundity to seed produced through outcross pollination ("outcrossed" seed). As a result, in most species there is strong natural selection to maximize the proportion of outcrossed seed (the "outcrossing rate").

Flowering plants are unusual among seed plants in their superlative exploitation of animals (primarily insects) as agents of outcross pollination. The outcross pollination efficiency of insects, birds, and mammals (primarily bats) may have contributed to both the abundance and diversity of flowering plants. Abundance may have increased because of less wastage of energy and resources on unsuccessful pollen and ovules. Diversity may have increased for two reasons. First, insects undoubtedly have selected for a wide variety of floral forms that provide different rewards (pollen and nectar) and are attractive in appearance (colour juxtaposition, size, shape) and scent (sweet, skunky) in different ways to different pollinators. Second, faithfulness of pollinators to particular familiar flowers may have reduced hybridization and speeded evolutionary divergence and the production of new species.

Although flowering plants first appeared after most of the major groups of insects had already evolved, flowering plants probably caused the evolution of many new species within these groups. Some new insect groups, such as bees and butterflies, originated after flowering plants, their members developing mouthpart structures and behaviour specialized for pollination. In extreme cases, a plant is completely dependent on one insect species for pollination, and the insect is completely dependent on one plant species for food. Such tight interdependency occurs rarely but is well documented in yuccas/yucca moths, senita cacti/senita moths. In all three insects, females lay eggs in the flowers, and their young hatch later to feed on the mature fruit and its contents. Females ensure that the fruit develops by gathering pollen from another plant and transporting it to the stigma of the flower holding their eggs. Plants benefit greatly in outcrossed seed produced, at the small cost of some consumed fruit and seeds, and the insects benefit greatly from the food supply for developing larvae at the small cost of transporting pollen the short distances between plants. Pollinating agents, whether biotic or abiotic, have exerted strong selection on all aspects of the flower, resulting in the evolution of tremendous floral diversity. This diversity has been distilled into a small number of characteristic pollination syndromes.

Pollination by beetles selects usually for white colour, a strong fruity scent, and a shallow, bowl-shaped flower. Bees select for yellow or blue/purple colorings, a landing platform with colour patterns that guide the bee to nectar (often reflecting in the ultraviolet range of the spectrum), bilateral symmetry, and a sweet scent. Butterflies select for many colors other than yellow, a corolla (petal) tube with nectar at the base, and the absence of any scent. Moths in contrast select for nocturnally opening flowers with a strong scent and drab or white colour, and also a tube with nectar at the base. Bats select also for nocturnally opening flowers, but with a strong musky scent and copious nectar, positioned well outside the foliage for easy access, and drab or white colour. Hummingbirds select for red or orange flowers with no scent, copious nectar production, and a corolla tube with nectar at the base. Other pollinating birds that do not hover while feeding select for strong perches and flowers capable of containing copious nectar (tubes, funnels, cup shapes).

Wind as a pollinating agent selects for lack of colour, scent, and nectar; small corolla; a large stigmatic surface area (usually feathery); abundantly produced, buoyant pollen; and usually erect styles and limp, hanging stamens. In addition there is great floral diversity within any of these syndromes, arising from the diverse evolutionary histories of the member plant species.

Selfing and Outcrossing

Most flowering plant species reproduce primarily by outcrossing, including the great majority of trees, shrubs, and perennial herbs. Adaptations that prevent self-fertilization include self-incompatibility (genetic recognition and blocking of self-pollen) and dioecy (separate male and female individuals). Adaptations that reduce the chances of self-pollination in hermaphrodites include separation of the anthers and stigma in space (herkogamy) or time (dichogamy). In many species, both self-incompatibility and spatiotemporal separation of the sex organs occur.

The ability to produce seeds by selfing, however, is advantageous in situations where outcrossing pollination is difficult or impossible. These include harsh environments where pollinators are rare or unpredictable, and regularly disturbed ground where survivors often end up isolated from each other. Selfing is also cheaper than outcrossing, because selfers can become pollinated without assistance from animals and therefore need not produce large, attractive flowers with abundant nectar and pollen rewards.

Most primarily selfing species are small annuals in variable or disturbed habitats, with small, drab flowers. Most desert annuals and roadside weeds, for example, are selfers. The evolutionary transition from outcrossing to near-complete selfing has occurred many times in flowering plants.

Outcrossing and selfing species differ in their evolutionary potential. Outcrossers are generally more genetically diverse and produce lineages that persist over long periods of evolutionary time, during which many new species are formed. Selfers, however, are less genetically diverse and tend to accumulate harmful mutations. They typically go extinct before they have an opportunity to evolve new species.

Asexual Reproduction

The ability to produce new individuals asexually is common in plants. Under appropriate experimental conditions, nearly every cell of a flowering plant is capable of regenerating the entire plant. In nature, new plants may be regenerated from leaves, stems, or roots that receive an appropriate stimulus and become separated from the parent plant. In most cases, these new plants arise from undifferentiated parenchyma cells, which develop into buds that produce roots and shoots before or after separating from the parent.

New plants can be produced from aboveground or belowground horizontal runners (stolons of strawberries, rhizomes of many grasses), tubers (potato, Jerusalem artichoke, dahlia), bulbs (onion, garlic), corms (crocus, gladiola), bulbils on the shoot (lily, many grasses), parenchyma cells in the leaves (Kalanchoe, African violet, jade plant) and inflorescence (arrowhead).

Vegetative propagation is an economically important means of replicating valuable agricultural plants, through cuttings, layering, and grafting. Vegetative reproduction is especially common in aquatic vascular plants (for example, surfgrass and eelgrass), from which fragments can break off, disperse in the current, and develop into new whole plants.

A minority of flowering plants can produce seeds without the fusion of egg and sperm (known as parthenocarpy or agamospermy). This occurs when meiosis in the ovule is interrupted, and a diploid egg cell is produced, which functions as a zygote without fertilization. Familiar examples include citrus, dandelion, hawkweed, buttercup, blackberry/raspberry, and sorbus. Agamospermous species are more common at high elevations and at high latitudes, and nearly all have experienced a doubling of their chromosome number (tetraploidy) in their recent evolutionary history. These species experience evolutionary advantages and disadvantages similar to those of selfers.

Evolutionary Significance of Plant Reproduction Strategies

The attractive, colorful, and unique features of the most abundant and diverse group of land plants—the flowering plants—are believed to have evolved primarily to maximize the efficiency and speed of outcross reproduction. Each major burst of angiosperm evolution was a coevolutionary episode with associated animals, primarily insects, which were exploited to disperse pollen and seeds in ever more efficient and diverse ways.

The first major burst of flowering plant evolution was the appearance of the closed carpel together with showy flowers that were radially symmetrical. The closed carpel prevented self-fertilization through recognition and blocking of self pollen within the specialized conducting tissue of the style. Insects attracted to the showy flowers carried pollen between plants less wastefully than wind, and the radial symmetry accommodated insects of many sizes and shapes.

The second major burst was the appearance of bilaterally symmetrical flowers, which happened independently in many groups of plants at the same time that bees evolved. Bilateral symmetry forced bees to enter and exit flowers more precisely, promoting even more efficient outcross pollen transfer.

The third major burst of flowering plant evolution was the appearance of nutritious, fleshy fruits and seeds, coincident with a diversification of birds and rodents. The exploitation of vertebrates for fruit and seed dispersal resulted in less haphazard transport of offspring to neighboring populations of the same species (also visited as a food source), thereby reducing the chances that progeny inbreed with their siblings and parents and providing more assurance than wind currents that they find good habitat and unrelated mating partners of the same species.

Sexual Reproduction

Sexual reproduction is characterized by processes that pass a combination of genetic material to offspring, resulting in increased genetic diversity. The main two processes are: meiosis, involving the halving of the number of chromosomes; and fertilization, involving the fusion of two gametes and the restoration of the original number of chromosomes. During meiosis, the chromosomes of each pair usually cross over to achieve homologous recombination.

The evolution of sexual reproduction is a major puzzle. The first fossilized evidence of sexually reproducing organisms is from eukaryotes of the Stenian period, about 1 to 1.2 billion years ago. Sexual reproduction is the primary method of reproduction for the vast majority of macroscopic organisms, including almost all animals and plants. Bacterial conjugation, the transfer of DNA between two bacteria, is often mistakenly confused with sexual reproduction, because the mechanics are similar. A major question is why sexual reproduction persists when parthenogenesis appears in some ways to be a superior form of reproduction. Contemporary evolutionary thought proposes some explanations. It may be due to selection pressure on the clade itself—the ability for a population to radiate more rapidly in response to a changing environment through sexual recombination than parthenogenesis allows. Alternatively, sexual reproduction may allow for the "ratcheting" of evolutionary speed as one clade competes with another for a limited resource. In the first stage of sexual reproduction, "meiosis," the number of chromosomes is reduced from a diploid number (2n) to a haploid number (n). During "fertilization," haploid gametes come together to form a diploid zygote and the original number of chromosomes (2n) is restored.

Plants

Animals typically produce male gametes called sperm, and female gametes called eggs and ova, following immediately after meiosis. With the gametes produced directly by meiosis. Plants on the other hand have mitosis occurring in spores, which are produced by meiosis. The spores germinate into the gametophyte phase. The gametophytes of different groups of plants vary in size; angiosperms have as few as three cells in pollen, and mosses and other so called primitive plants may have several million cells. Plants have an alternation of generations where the sporophyte phase is succeeded by the gametophyte phase. The sporophyte phase produces spores within the sporangium by meiosis.

Flowering Plants

Flowering plants are the dominant plant form on land and they reproduce by sexual and asexual means. Often their most distinguishing feature is their reproductive organs, commonly called flowers. The anther produces male gametophytes, the sperm is produced in pollen grains, which attach to the stigma on top of a carpel, in which the female gametophytes (inside ovules) are located. After the pollen tube grows through the carpel's style, the sex cell nuclei from the pollen grain migrate into the ovule to fertilize the egg cell and endosperm nuclei within the female gametophyte in a process termed double fertilization. The resulting zygote develops into an embryo, while the triploid endosperm (one sperm cell plus two female cells) and female tissues of the ovule give rise to the surrounding tissues in the developing seed. The ovary, which produced the female gametophyte(s), then grows into a fruit, which surrounds the seed(s). Plants may either self-pollinate or cross-pollinate. Nonflowering plants like ferns, moss and liverworts use other means of sexual reproduction.

Ferns

Ferns typically produce large diploid sporophytes with rhizomes, roots and leaves; and on fertile leaves called sporangium, spores are produced. The spores are released and germinate to produce short, thin gametophytes that are typically heart shaped, small and green in colour.

The gametophytes or thallus, produce both motile sperm in the antheridia and egg cells in separate archegonia. After rains or when dew deposits a film of water, the motile sperm are splashed away from the antheridia, which are normally produced on the top side of the thallus, and swim in the film of water to the antheridia where they fertilize the egg. To promote out crossing or cross fertilization the sperm are released before the eggs are receptive of the sperm, making it more likely that the sperm will fertilize the eggs of different thallus. A zygote is formed after fertilization, which grows into a new sporophytic plant. The condition of having separate sporephyte and gametophyte plants is called alternation of generations. Other plants with similar reproductive means include the *Psilotum*, *Lycopodium*, *Selaginella* and *Equisetum*.

Bryophytes

The bryophytes, which include liverworts, hornworts and mosses, reproduce both sexually and vegetatively. They are small plants found growing in moist locations and like ferns, have motile sperm with flagella and need water to facilitate sexual reproduction. These plants start as a haploid spore that grows into the dominate form, which is a multicellular haploid body with leaf-like structures that photosynthesize.

Haploid gametes are produced in antherida and archegonia by mitosis. The sperm released from the antherida respond to chemicals released by ripe archegonia and swim to them in a film of water and fertilize the egg cells thus producing a zygote. The zygote divides by mitotic division and grows into a sporophyte that is diploid. The multicellular diploid sporophyte produces structures called spore capsules, which are connected by seta to the archegonia. The spore capsules produce spores by meiosis, when ripe the capsules burst open and the spores are released. Bryophytes show considerable variation in their breeding structures and the above is a basic outline. Also in some species each plant is one sex while other species produce both sexes on the same plant.

Fungi

Fungi are classified by the methods of sexual reproduction they employ. The outcome of sexual reproduction most often is the production of resting spores that are used to survive inclement times and to spread. There are typically three phases in the sexual reproduction of fungi: plasmogamy, karyogamy and meiosis.

Insects

Insect species make up more than two-thirds of all extant animal species, and most insect species use sex for reproduction, though some species are facultatively parthenogenetic. Many species have sexual dimorphism, while in others the sexes look nearly identical. Typically they have two sexes with males producing spermatozoa and females ova. The ova develop into eggs that have a covering called the chorion, which forms before internal fertilization. Insects have very diverse mating and reproductive strategies most often resulting in the male depositing spermatophore within the female, which stores the sperm until she is ready for egg fertilization. After fertilization, and the formation of a zygote, and varying degrees of development; the eggs are deposited outside the female in many species, or in some, they develop further within the female and live born offspring are produced.

Mammals

There are three extant kinds of mammals: Monotremes, Placentals and Marsupials, all with internal fertilisation. In placental mammals, offspring are born as juveniles: complete animals with the sex organs present although not reproductively functional. After several months or years, the sex organs develop further to maturity and the animal becomes sexually mature. Most female mammals are only fertile during certain periods during their estrous cycle, at which point they are ready to mate. Individual male and female mammals meet and carry out copulation. For most mammals, males and females exchange sexual partners throughout their adult lives.

Male

The male reproductive system contains two main divisions: the penis, and the testicles, the latter of which is where sperm are produced. In humans, both of these organs are outside the abdominal cavity, but they can be primarily housed within the abdomen in other animals (for instance, in dogs, the penis is internal except when mating). Having the testicles outside the abdomen best facilitates temperature regulation of the sperm, which require specific temperatures to survive. Sperm are the smaller of the two gametes and are generally very short-lived, requiring males to produce them continuously from the time of sexual maturity until death. Prior to ejaculation the produced sperm are stored in the seminal vesicle, a small gland that is located just behind the bladder.

Female

The female reproductive system likewise contains two main divisions: the vagina and uterus, which act as the receptacle for the sperm, and the ovaries, which produce the female's ova. All of these parts are always internal. The vagina is attached to the uterus through the cervix, while the uterus is attached to the ovaries via the Fallopian tubes. At certain intervals, the ovaries release an ovum, which passes through the fallopian tube into the uterus.

If, in this transit, it meets with sperm, the sperm penetrate and merge with the egg, fertilizing it. The fertilization usually occurs in the oviducts, but can happen in the uterus itself. The zygote then implants itself in the wall of the uterus, where it begins the processes of embryogenesis and morphogenesis. When developed enough to survive outside the womb, the cervix dilates and contractions of the uterus propel the fetus through the birth canal, which is the vagina.

The ova, which are the female sex cells, are much larger than the sperm and are normally formed within the ovaries of the fetus before its birth. They are mostly fixed in location within the ovary until their transit to the uterus, and contain nutrients for the later zygote and embryo. Over a regular interval, in response to hormonal signals, a process of oogenesis matures one ovum which is released and sent down the Fallopian tube. If not fertilized, this egg is flushed out of the system through menstruation in humans and other great apes and reabsorbed in other mammals in the estrus cycle.

Gestation

Gestation, called *pregnancy* in humans, is the period of time during which the fetus develops, dividing via mitosis inside the female. During this time, the fetus receives all of its nutrition and oxygenated blood from the female, filtered through the placenta, which is attached to the fetus' abdomen via an umbilical cord. This drain of nutrients can be quite taxing on the female, who is required to ingest slightly higher levels of calories. In addition, certain vitamins and other nutrients are required in greater quantities than normal, often creating abnormal eating habits. The length of gestation, called the gestation period, varies greatly from species to species; it is 40 weeks in humans, 56–60 in giraffes and 16 days in hamsters.

Birth

Once the fetus is sufficiently developed, chemical signals start the process of birth, which begins with contractions of the uterus and the dilation of the cervix. The fetus then descends to the cervix, where it is pushed out into the vagina, and eventually out of the female. The newborn, which is called an infant in humans, should typically begin respiration on its own shortly after birth. Not long after, the placenta is passed as well. Most mammals eat this, as it is a good source of protein and other vital nutrients needed for caring for the young. The end of the umbilical cord attached to the young's abdomen eventually falls off on its own.

Monotremes

Monotremes, only five species of which exist, all from Australia and New Guinea, lay eggs. They have one opening for excretion and reproduction called the cloaca. They hold the eggs internally for several weeks, providing nutrients, and then lay them and cover them like birds. After less than two weeks the young hatches and crawls into its mother's pouch, much like marsupials, where it nurses for several weeks as it grows.

Marsupials

Marsupials reproductive systems differ markedly from those of placental mammals. Females have two vaginas, both of which open externally through one orifice but lead to different compartments within the uterus. Males generally have a two-pronged penis, which corresponds to the females' two vaginae. The penis is used only for discharging semen into females, and is separate from the urinary tract. Both sexes possess a cloaca, which is connected to a urogenital sac used to store waste before expulsion. The female develops a kind of yolk sack in her womb which delivers nutrients to the embryo. Embryos of bandicoots, koalas and wombats additionally form placenta-like organs that connect them to the uterine wall, although the placenta-like organs are smaller than in placental mammals and it is not certain that they transfer nutrients from the mother to the embryo. Pregnancy is very short, typically 4 to 5 weeks. The embryo is born at a very young stage of development, and is usually less than 5 cm

(2.0 in) long at birth. It has been suggested that the short pregnancy is necessary to reduce the risk that the mother's immune system will attack the embryo. The newborn marsupial uses its forelimbs (with relatively strong hands) to climb to a nipple, which is usually in a pouch on the mother's belly. The mother feeds the baby by contracting muscles over her mammary glands, as the baby is too weak to suckle. The newborn marsupial's need to use its forelimbs for climbing to the nipple has prevented the forelimbs from evolving into paddles or wings and has therefore prevented the appearance of aquatic or truly flying marsupials (although there are several marsupial gliders).

Fish

The vast majority of fish species lay eggs that are then fertilized by the male, some species lay their eggs on a substrate like a rock or on plants, while others scatter their eggs and the eggs are fertilized as they drift or sink in the water column. Some fish species use internal fertilization and then disperse the developing eggs or give birth to live offspring. Fish that have live-bearing offspring include the Guppy and Mollies or *Poecilia*. Fishes that give birth to live young can be ovoviviparous, where the eggs are fertilized within the female and the eggs simply hatch within the female body, or in seahorses, the male carries the developing young within a pouch, and gives birth to live young. Fishes can also be viviparous, where the female supplies nourishment to the internally growing offspring. Some fish are hermaphrodites, where a single fish is both male and female and can produce eggs and sperm. In hermaphroditic fish, some are male and female at the same time while in other fish they are serially hermaphroditic; starting as one sex and changing to the other. In at least one hermaphroditic species, self-fertilization occurs when the eggs and sperm are released together. Internal self-fertilization may occur in some other species. One fish species does not reproduce by sexual reproduction but uses sex to produce offspring; *Poecilia formosa* is a unisex species that uses a form of parthenogenesis called gynogenesis, where unfertilized eggs develop into embryos that produce female offspring. *Poecilia formosa* mate with males of other fish species that use internal fertilization, the sperm does not fertilize the eggs but stimulates the growth of the eggs which develops into embryos.

Plant Reproduction

The plants that sexually reproduce have the reproductive structures called the flowers. The flower is a condensed shoot with the nodes present very close to each other. The different parts of the plant are attached to the nodes. All the structures present at one node are collectively called the whorl. The first or the outermost two whorls are called the non-reproductive whorls. They are the calyx and corolla. The inner two whorls are androecium and gynoecium, the reproductive whorls.

Parts of a Flower

Calyx: is the outermost and most often green in colour. The individual units of calyx are called the sepals. It protects the inner whorls at bud stage.

Corolla: is the next inner whorl and is often coloured brightly. The individual units of corolla are called petals. They serve to attract bees, birds, etc which are the agents of pollination. Androecium is the male reproductive part of the flower. The individual units of androecium are called the stamens. Each stamen has a thread-like filament at the free end of which is attached the four-lobed anther.

Structure of Anther

The anther has four pollen-sacs, one in each lobe. The pollen-sacs contain cells called the microspore mother cells (MMC) or the pollen mother cells. MMCs undergo mieosis to produce four haploid microspores, also called the pollen grains. Each microspore divides once mitotically to produce two male gametes or sperm cells. Thus, each microspore mother cell produces 8 sperm cells. Gynoecium is the female reproductive part of the flower. The individual units are called the carpels or pistils.

A flower may have one to many carpels, either fused or free. Each carpel is made up of the basal ovary, middle style and the upper stigma. The ovary is the chamber where there are many ovules that are attached to an axis. Each ovule consists of a haploid egg and other associated cells. The stigma is a sticky structure that receives the pollen grains. The style is hollow and provides a passage for the male gametes to reach the female gametes, the eggs.

Pollination

Transfer of pollen grains to the stigma is called pollination. If the pollen grains are transferred to the stigma of the same flower, the pollination is called self-pollination or autogamy. If the pollen grains are transferred to the stigma of another flower of the same species, the pollination is called cross-pollination or allogamy.

The anthers on maturity burst open with force and this is called dehiscence. This releases the pollen grains with force which are then carried by wind and water to other plants. In other plants, the flowers are brightly coloured and scented to attract the birds, bees, etc. The insect or the bird enters the flower to suck the nectar produced by glands at the base of the flower. The pollen grains present on the dehisced or open anther, stick to the legs or abdomen. When the same insect visits other flowers the pollen grains are transferred to the stigma of those flowers.

Fertilisation

The tip of the tube contains the male nuclei. The tube grows and enters the ovule where it bursts at the tip releasing the male gametes. One of the male gametes fuses with the egg, the female gamete. The fusion of the male gamete with the female gamete is called fertilisation. This results in the formation of zygote that is diploid. The zygote develops into the embryo. The other male gamete fuses with the polar nuclei. This results in the formation of a triploid nucleus called the endosperm nucleus. Since the process of fertilisation involves two fusions, it is called double fertilisation.

The divisions of the endosperm nucleus result in the formation of the endosperm that nourishes the growing embryo. The ovule then becomes the seed and the ovary changes into fruit.

Plant Reproduction

The way plants reproduce is a totally fascinating process. Out of the more than 300,000 different kinds of plants, more than half are seed plants. Seed plants make their own seeds from which new plants grow. Other ways plants can make new plants are from spores, rhizomes, bulbs, tubers, corms, cuttings, grafts, and buds. Seeds are made by flowers in some plants and by cones in other plants. Flower parts contain specialized cells, called ovules, which have the job of making the seeds. Different kinds of flowers have many different parts. Basically, most plants—trees, vegetables, and even grass—sprout flowers. Some of these flowers are beautiful and colorful while others are so tiny one can hardly see them, but all plants produce new plants.

Botanists classify the reproduction of plants as ones with perfect and imperfect flowers. Perfect flowers are the ones that contain both the male and female parts. Imperfect flowers are the ones that contain only male or only female parts. So, of course, these different flowers have to find different ways to propagate themselves.

The stamen is the male part of a flower and has an anther on a stalk. The anther's job is to produce the pollen. Then there is the pistil, or female part. The pistil contains the flower's ovary, the style, and the stigma. Inside the ovary are the ovules. Each ovule contains an egg cell. So when an egg cell joins with a pollen cell, a seed may be produced. The ovary is at the base of the plant. The stigma is at the top, and between them is the style. So, the pollen from the anther has to land on the stigma. When that happens, the pollen germinates and makes its way down into the ovary. Inside the ovary are the ovules, which contain egg cells. When an egg cell joins with a pollen cell, a seed may be produced. Then the ovary becomes a fruit. So a fruit is a mature ovary holding the seeds. In fruits, like the pears, the pear's core is the seedpod of the pear tree. Acorns are the seedpods of oak trees. And the tomato is the seedpod of the tomato plant.

A flower garden is alive with insects. They are all after the sweet liquid called nectar, which the flowers produce, near the bottom of the pistil. Flowers produce the nectar to draw the insects to them. Insects—like butterflies, wasps, and bees—come to get the nectar. Some, like butterflies and wasps, use the nectar for food; some, like bees, make it into honey and store it in their hives. If you look closely at the body of a butterfly or bee, you will notice it is covered with tiny hairs. As they land on the flower to collect the nectar, some of the pollen is brushed on these hairs. Then the insect carries the pollen to the next flower, and pollination takes place. Some plants, like corn, or flowers of an oak tree, rely on the wind to spread their pollen. So these kinds of plants don't produce pretty flowers since they don't need to attract insects to come and pollinate them.

Some plants need long tongues to pollinate them. Their flowers are usually tubular, like the trumpet creeper vine. These kinds of flowers are best pollinated by hummingbirds or by moths that have long tongues. And flowers that open at night do so because they need the specialized pollination provided by certain moths. Even the colour of plants makes a difference. Birds, like the hummingbird, usually pollinate plants that are red. Night insects pollinate white and pale yellow flowers, which are seen better at night.

Trees that have needle-like leaves are called conifers. Conifers produce cones instead of flowers. Under each scale of a cone is a part that produces either pollen or an ovule. But a cone cannot produce both. The very small cones produce the pollen in the spring. The wind carries the pollen to the larger cone, which has the ovule producing part. Then, the pollen producing cones fall off while the fertilized ovules grow larger. Eventually, they open and release the seeds, and some of the seeds sprout. Some plants, like tulips, grow from bulbs. Bulbs produce plants very easily by reproducing underground.

Plants can be grown without seeds by cuttings. A cutting can be from any part of the fully-grown plant, and when it is placed in soil, it may produce a new plant. You can increase your own houseplants by this method, and here is the way to do it. With a sharp knife, cut a piece of stem from the houseplant you want to use.

Pull off any leaves or flowers from the stem's lower part. Stick the stem in some clean, moist sand. Cover the flowerpot with a clear, plastic bag. Place in a well-lit place. After a few weeks, lift the stem out of the sand. Are the roots beginning to grow? Now you can plant it in a pot with potting soil. Keep the soil moist, and soon, the new plant will begin to grow. When cuttings from trees and other woody stems are attached to another plant, this is known as grafting. Grafts are another way of growing plants without using seeds.

There are a variety of ways plants produce new plants. To find out more about this fascinating process, you may want to read books on the subject or do some further research on the Internet.

Genetic Variations in Plants

Variation

If you have grown Fast Plants® you will have noticed variation among the individual plants within the group. Variation can range from a little to a lot. This article is designed to help teachers and students understand the basis of the variation they observe in their plantings.

Variation is one of the fundamental characteristics of life. All organisms exhibit some variation among individuals. Understanding the ways that variation is manifested in organisms, how it comes to be expressed through the development of the individual, and how it is transmitted from one individual to the next generation of individuals are central themes in biology.

Working with Fast Plants® will enrich a student's understanding of variation. By observing the growth and development of a Fast Plants® through the various stages in the life cycle, students will become aware of many visible features, or phenotypes, that make up the organism.

Only, however, upon close observation of a population of plants, will they become aware that the characteristics observed on one plant vary more or less on other plants. Such is the nature of variation.

Phenotypic Variation

Phenotypic variation, e.g. plant height at a particular stage of development, is considered to be the expression of the genetic makeup (genotype) of the individual as it interacts with the environment. Variation in plant height among individuals in a population is therefore due to variation in the interaction between the genotype and the environment, as expressed through the development of each individual plant.

Describing and Observing Phenotypic Variation

In order to be useful in an experiment the phenotype must be described using terms that are widely understood and easily communicated. For these reasons scientists have agreed upon various standards or descriptors to describe characteristics in the natural world. Descriptors take many forms, see WFPID Observing and Describing. The choice of how to describe what you observe is important, because it will determine the kinds of descriptors used and establish the basis for recording, analyzing and communicating results.

Environmental Variation

Much can be learned about the role of light, temperature, and nutrients on plant development from experiments in which one or more environmental parameters are changed. Because Fast Plants® are highly responsive to changes in the environment, they are ideal for examining the role of the environment on the expression of phenotypic variation. Although Fast Plants® are able to grow within a wide range of environmental conditions, for most investigations it is recommended that they be grown under uniform and ideal conditions. In this way variation arising from sub-optimal conditions of environment will be minimized. The Wisconsin Fast Plants® Information Document (WFPID) Understanding the Environment describes how to provide and maintain the various physical, chemical and biotic components of the environment that are most suitable for Fast Plants®.

Genotypic Variation

How can students use Fast Plants® to investigate the contribution of the genotype to the phenotype? Fast Plants®, rapid cycling Brassica rapa, are genotypically variable in that they have a genetically controlled mating system that prevents self-fertilization and favours out-crossing among individuals. As a consequence, even seed stocks selected for uniformity of specific phenotypes and genotypes exhibit considerable variation for other traits.

The Wisconsin Fast Plants® Program has developed a number of genetic stocks of rapid cycling Brassica rapa for genetic investigations on the nature and inheritance of variation. Some WFP stocks contain distinctive mutant phenotypes, e.g., anthocyaninless plant, anl, yellow green plant, ygr1, rosette, ros, and male sterile, mst2, that are suitable for Mendelian genetics. Other stocks exhibit phenotypes whose expression may vary continually and which may be quantified as discrete or countable units, e.g., number of hairs, or as estimates of size, e.g. petite dwarf, dwf1, or of intensity of colour saturation, e.g., purple anthocyanin. These quantitative phenotypes may be conditioned by a few or many genes and normally require numerical descriptions in which the statistical notations of population size, (n), range (r), arithmetic mean (x), and standard deviation (s) are applied.

Yet other stocks have been developed to combine both simply inherited mutant genotypes and quantitatively expressed phenotypes. An important part of WFP is the continuing development and improvement of seed stocks for uses in genetics.

Genetic Diversity

Genetic diversity is a level of biodiversity that refers to the total number of genetic characteristics in the genetic makeup of a species. It is distinguished from genetic variability, which describes the tendency of genetic characteristics to vary.

The academic field of population genetics includes several hypotheses and theories regarding genetic diversity. The neutral theory of evolution proposes that diversity is the result of the accumulation of neutral substitutions. Diversifying selection is the hypothesis that two subpopulations of a species live in different environments that select for different alleles at a particular locus. This may occur, for instance, if a species has a large range relative to the mobility of individuals within it. Frequency-dependent selection is the hypothesis that as alleles become more common, they become less fit. This is often invoked in host-pathogen interactions, where a high frequency of a defensive allele among the host means that it is more likely that a pathogen will spread if it is able to overcome that allele.

Importance of Genetic Diversity

There are many different ways to measure genetic diversity. The modern causes for the loss of animal genetic diversity have also been studied and identified. A 2007 study conducted by the National Science Foundation found that genetic diversity and biodiversity are dependent upon each other — that diversity within a species is necessary to maintain diversity among species, and vice versa. According to the lead researcher in the study, Dr. Richard Lankau, "If any one type is removed from the system, the cycle can break down, and the community becomes dominated by a single species."

Survival and Adaptation

Genetic diversity plays a very important role in survival and adaptability of a species because when a species's environment changes, slight gene variations are necessary to produce changes in the organisms' anatomy that enables it to adapt and survive. A species that has a large degree of genetic diversity among its population will have more variations from which to choose the most fit alleles. Increase in genetic diversity is also essential for an organism to evolve. Species that have very little genetic variation are at a great risk. With very little gene variation within the species, healthy reproduction becomes increasingly difficult, and offspring often deal with similar problems to those of inbreeding. The vulnerability of a population to certain types of diseases can also increase with reduction in genetic diversity.

Agricultural Relevance

When humans initially started farming, they used selective breeding to pass on desirable traits of the crops while omitting the undesirable ones. Selective breeding leads to monocultures: entire farms of nearly genetically identical plants. Little to no genetic diversity makes crops extremely susceptible to widespread disease. Bacteria morph and change constantly. When a disease causing bacterium changes to attack a specific genetic variation, it can easily wipe out vast quantities of the species. If the genetic variation that the bacterium is best at attacking happens to be that which humans have selectively bred to use for harvest, the entire crop will be wiped out.

A very similar occurrence is the cause of the infamous Potato Famine in Ireland. Since new potato plants do not come as a result of reproduction but rather from pieces of the parent plant, no genetic diversity is developed, and the entire crop is essentially a clone of one potato, it is especially susceptible to an epidemic. In the 1840s, much of Ireland's population depended on potatoes for food. They planted namely the "lumper" variety of potato, which was susceptible to a rot-causing plasmodiophorid called *Phytophthora infestans*. This plasmodiophorid destroyed the vast majority of the potato crop, and left tens of thousands of people to starve to death.

Coping with Poor Genetic Diversity

The natural world has several ways of preserving or increasing genetic diversity. Among oceanic plankton, viruses aid in the genetic shifting process. Ocean viruses, which infect the plankton, carry genes of other organisms in addition their own. When a virus containing the genes of one cell infects another, the genetic makeup of the latter changes. This constant shift of genetic make-up helps to maintain a healthy population of plankton despite complex and unpredictable environmental changes.

Cheetahs are a threatened species. Extremely low genetic diversity and resulting poor sperm quality has made breeding and survivorship difficult for cheetahs -- only about 5% of cheetahs survive to adulthood. About 10,000 years ago, all but the jubatus species of cheetahs died out. The species encountered a population bottleneck and close family relatives were forced to mate with each other, or inbreed. However, it has been recently discovered that female cheetahs can mate with more than one male per litter of cubs. They undergo induced ovulation, which means that a new egg is produced every time a female mates. By mating with multiple males, the mother increases the genetic diversity within a single litter of cubs.

Measures of Genetic Diversity

Genetic Diversity of a population can be assessed by some simple measures.

- Gene Diversity is the proportion of polymorphic loci across the genome.
- Heterozygosity is the mean number of individuals with polymorphic loci.
- Alleles per locus is also used to demonstrate variability.

The Quest for the Hardy Cedar-of-Lebanon

Genetic variation within plant species has not only been richly documented in science but also widely exploited for horticultural use. Individual plants have often been selected from wild populations for their deviations in growth habit, flower size, and leaf colour, but another primary driver of plant exploration has been the promise of winter har-diness due to provenance. In many cases, this, too, is under the auspices of science, for it allows botanical gardens and arboreta to cultivate species that may normally be out of reach due to lack of hardiness. We present here a prime example of such work: the quest by the Arnold Arboretum to introduce into cultivation hardy stock of the cedar-of-Lebanon, Cedrus libani.

The genus Cedrus has a natural range that extends from North Africa around the Medi-terranean Sea into Lebanon, Syria, Cyprus, and Turkey and west to the Himalayas (Far-jon 1990). Depending on the treatment, there are a variable number of species of true cedar, and Farjon (2001) recognizes four. The beauti-ful long-needled deodar cedar (Cedrus deodora) occurs in a wide range of habitats in the Hima-layas of Afghanistan, Pakistan, Kashmir, and Nepal. The atlas cedar (Cedrus atlantica) grows in the Atlas Mountains of Algeria and Morocco while the short-needled Cyprus cedar (Cedrus brevifolia) is restricted to that island; both of these taxa have by some botanists been consid-ered separate subspecies of Cedrus libani, the cedar-of-Lebanon, which occurs naturally in Lebanon, Syria, and the Cilician Taurus moun-tains of southern Anatolia, or modern-day Turkey. It is this group of cedars from Turkey that most interests us and that is the focus of this article.

Authors have variously recognized the Turk-ish provenance of cedar with subspecific status (Farjon 1990). Cedrus libani subspecies steno-coma was first described by Schwarz (1944) and then Davis (1949), who both recognized that it was intermediate between the typical cedar-of-Lebanon and the atlas cedar. Volume One of The Flora of Turkey (Davis 1965) did not rec-ognize subspecies stenocoma, although it is recognized as a variety in the eleventh volume of the same work (Guner 2000). More recently, the Turkish provenance was classified as Cedrus libani ssp. stenocoma (Farjon 2001) and these trees usually are called the hardy cedar-of-Lebanon. The Turkish trees are generally considered to be more upright and conical

Cedrus Libani in Asia Minor

During the 1800s Cedrus libani was grown throughout Philadelphia and New York but was not hardy in Boston and New England (Wilson 1926). Josiah Hoopes, a nurseryman from West Chester, in southeastern Pennsylvania, wrote that "the cedar-of-

Lebanon is found to be pretty hardy... [and] with us it has succeeded to our entire satisfaction, and we can therefor recommend it without reserve, if proper cultivation and a moderate amount of care be given to it." In his 1868 book, Hoopes also mentions Pierre Belon, a French botanist and physician who travelled throughout the Levant in the 1540s. In 1553 Belon published De arboribus coniferis, probably the first text devoted entirely to conifers, and in it he included the first description of cedar-of-Lebanon growing in the Amanus (Nur) and Taurus Mountains of southern Anatolia. In 1597, John Gerard cited Belon in his Herbal, saying:

Gerard's statement reveals that as early as the sixteenth century authors took notice of these unique trees from Anatolia and recognized that this more northern provenance possessed greater potential for cold hardiness.

Nineteenth-century botanical and horticultural literature was replete with references to the northern populations of cedar-of-Lebanon. Several European botanists were exploring and describing the flora of Asia Minor and their works describe a growing understanding of the natural range of cedar in this region. Asie Mineure by P. A. Chikhachev is an extremely thorough account of the physical geography, climate, fauna, and flora of Asia Minor, based on his travels throughout the region. Among his extensive botanical listings is Cedrus libani, which he described as growing in numerous locations. Chikhachev describes their locations using the ancient names for the regions of Anatolia, moving from west to east: in Pisidia between Lakes Beysehir and Egridir; in Isauria at Mount Topyedik growing around 2,000 meters; in Cilicia growing on the northeast and southeast exposures of all the mountainous regions of the Bulgar Daglari (Bolkar Daglari mountains of the Taurus range), where it descends to 4,600 feet (1,400 m); growing in groves in the Antitaurus Mountains between the villages Sarkanty-oglu and Tchedeme [sic] at 5,600 feet (1,700 m) (Chikhachev 1860).

The nearest point to the Lebanon [Mountains] at which Cedars have been found, is the Bulgar-dagh chain of the Taurus in Asia Minor, and from that point forests extend eastward to Pisidia, in long. E. 32°, westward to long. E. 36°, and north-ward to the Anti-Taurus, in lat. 40° N.; growing at elevations of 4000 to 6400 feet above the sea. The Lebanon may be regarded as a branch of the Taurus, and is 250 miles distant from the Cedar forests upon that chain... Northern Syria and Asia Minor form one botanical province; so that the Lebanon grove, though so widely disconnected from the Taurus forests, can be regarded in no other light than as an outlying member of the latter. Ravenscroft gave a colossal summary of all known accounts of cedar-of-Lebanon, with beautiful colour plates, in his 1884 Pinetum Bri-tannicum. He describes the species in amazing detail, including comparisons made to other Cedrus and descriptions of the Syrian and Leba-nese trees. Particularly interesting is a table that accounts for all visits made by individuals from 1487 to 1864 to the sacred grove that lies on Mt. Lebanon. He also writes of the Anatolian population, providing a description practically identical to that of Hooker.

How the Hardy Cedar-of-Lebanon Found Its Way to North America

A contemporary of the botanists writing and exploring Asia Minor in the late 1800s was Walter Siehe, an interesting and somewhat mysterious character who played an integral role in introducing the hardy cedar-of-Lebanon into the United States. Siehe was a German botanical explorer living in Smyrna (Izmar), Turkey. In his Die Nadelholzer des cilicischen Taurus (Conifers of the Cilician Taurus), he described the natural habitat of conifers growing in the mountains of southern Anatolia and the conifers themselves, including Abies cilicica, Taxus baccata, several species of Juniperus, Cupressus sempervirens, and of course Cedrus libani. Of the cedar he wrote, “the proud tree is a child of the high altitude,” growing in a severe cli-mate where the snows lie a few meters deep for five months of the year (Siehe 1897b).

In these mountains where the cedars grow on the steep walls and high saddles between peaks, Siehe romantically describes the roaring wild rivers, the whiteness of the snow, and the long silences broken only by the screech of birds or tumbling stones kicked loose by an escaping mountain goat. He notes that despite the usefulness of cedar's wood, the large populations of cedar persisted because of the inaccessibility of the mountains. He describes trees up to 130 feet (40 m) tall and describes their column-like trunks supporting branches as regular as floors of a building (Siehe 1897a). It is clear from his travel accounts that Siehe knew the mountains of southern Turkey very well and was the right person, in the right place, and at the right time to send seed to North America.

It is in the context of nineteenth-century botanical exploration and description through-out Asia Minor that one understands Charles S. Sargent's interest in the more northern population of cedar. Cedar-of-Lebanon—with its handsome and stately form, its association with grand estates throughout Europe and the mid-Atlantic United States, and its historic asso-ciations—is a highly desirable landscape tree. With this in mind, it's easy to understand how Sargent, director of the Arnold Arboretum from 1873 until 1927, would have surely longed to grow cedars in New England. However, hardiness in New England was indeed an issue. He certainly would have read with great interest the accounts of the cedars and understood that trees from the Taurus mountains held the key to increased hardiness. With this in mind, Sargent hired Walter Siehe to collect seed from trees in the Taurus Mountains and have these sent to the Arnold Arboretum. In a letter from Siehe to Sargent, dated 18 November 1900, from Mersina, Turkey, Siehe wrote:

Dr. Bolle... has repeatedly informed me of your desire [to acquire] cedar cones from cold resistant trees of high altitude (1900 m [6,250 feet]). Only a few days ago did I manage to obtain, after several futile attempts, 50 kilos [110 pounds] of cones with good seeds. Since it was necessary to make a special trip, use many pack animals, and spend eight days of time for this, I am certain that you will not find the fee of 60 Mark German currency too high.

Apparently Sargent did not find the fee too high because the Arboretum's plant records show that they received cones with ripe seeds from Siehe on February 4, 1902. The Arnold Arboretum was not the only recipient of Siehe's seed: in 1908 H. J. Elwes and Augustine Henry wrote another excellent description of the cedars from the Taurus mountains and noted, "Siehe has sent seed from the Cilician Taurus to various places, and I have two vigorous young trees raised from them."

Early reports from the Arnold Arboretum noted great success with this seedlot. The seeds had a high rate of germination and by 1915, Sargent reported in the Arboretum's Bulletin of Popular of Information that the cedars-of-Lebanon had "all proved perfectly hardy, not one having suffered from drought or cold." A plantation of trees was established on Bussey Hill, and other individual specimens were planted throughout the collection. The average height of these trees was about 13 feet (4 m), with the tallest having reached 21 feet (6.5 m), prompting Sargent to reflect, "It is doubtful if any other conifer can be grown in New England from seed to the height of twenty-one feet in thirteen years." Wilson also seemed pleased with the rapid growth and hardiness of these trees, writing in 1919 that although the dreadful winter of 1917–1918 scorched the needles of the cedars, they recovered fully and "had grown more rapidly in the Arnold Arboretum than any other conifer has ever done." And just five years later, the Bulletin reported that the trees had already reached 30 feet (9 m) in height. In 1926, twenty individuals of this accession (AA 4697*A-T) appear in the plant records, although over time there has been some attrition and currently only eight trees from Siehe's original 1902 collection remain extant in the Arnold's collection (AA #4697*A,C,G,I,K,M,O,P). Cold winters were not the cause for their decline, however; Donald Wyman wrote in 1946 that the cedars were thriving, growing for over forty years and withstanding temperatures of minus-20 degrees F. Strong winds were responsible for the loss of at least eight of the twelve trees, including five from the infamous 1938 hurricane alone.

The Current State of Trees in the Wild

In Lebanon and Syria, the species is rare due to millennia of human impact (logging, burning, grazing). However, in Turkey, where the topography has prevented easy access, there remain extensive forests of Cedrus libani ssp. stenocoma. During a collecting expedition to the Taurus Mountains in 1990, Mark Flanagan encountered the hardy cedar-of-Lebanon running for nearly 620 miles (1,000 km) along the 5,900-foot (1800-m) contour line. Due to the high, open canopies in the overstory, the ground layer of these forests is very rich, including a diversity of taxa such as Acer hyrcanum, Sorbus umbellata, and Kitaibelia balansae.

Other Notably Hardy Specimens

Among the Morris Arboretum's extensive conifer collection are five mature specimens of hardy cedar-of-Lebanon. They are growing throughout the Arboretum, some tucked away, others in full view, and all of them handsome.

The oldest tree was planted before the Arboretum's founding in 1932 when the property was known as Compton, John and Lydia Morris's estate. A cedar appears in this location in the 1909 Atlas of Compton—a survey of the Morris' gardens and plants—and according to notes written by John Tonkin, the Morris' gardener and Morris Arboretum's superintendent from 1913 to 1961, this cedar came from the Arnold Arboretum. It is likely that this tree is a seed-ling from the original 1902 collection of Turk-ish seed, sent to the Morris's from the Arnold Arboretum. Over the years, staff at the Morris Arboretum have marvelled at the hardiness of this tree, recording that it showed no visible injuries during the devastatingly cold winters in the early 1930s (Lambert 1936). Although it has suffered storm damage during the past 25 years, today it shows a remarkable amount of young, vigorous re-growth for a tree of its age and size.

As students of Harrison Flint at Purdue University we often admired the selection 'Purdue Hardy' (Cedrus libani ssp. stenocoma 'Purdue Hardy'), which grows in West Lafayette, Indiana, in a hardiness zone that routinely reaches minus-20 degrees F (USDA zone 5a). This 40-year-old specimen is remarkable for its graceful form and nearly pendant branches; like many others of the subspecies, it has not become flat-topped.

It has certainly lived up to its name, withstanding winter temperatures of minus-25 degrees F with only minimal browning of needles (Flint 1997). The selection's provenance is uncertain. It was one of several seedlings germinated from seed collected by the late Purdue professor Ted Shaw in the 1950s. Shaw had been in Lebanon working on reforestation projects supported by the United States when he obtained it.

Oral history at Purdue has it that Shaw found the seeds "up in the Hills," which could mean Lebanon or it could have been Turkey, where he vacationed.

Since no Lebanese cedar has been successfully grown out-of-doors north of zone 6, it is far more likely to have originated in the mountains of Turkey. Another noteworthy specimen, a mammoth cedar-of-Lebanon that is a Pennsylvania state champion, is at the Tyler Arboretum, in Media, Pennsylvania.

Jacob and Minshall Painter, horticulturists and owners of the property that became the Tyler Arboretum, recorded purchases of cedars of Lebanon from the Phila-delphia nurseries of John Evans, Josiah Hoopes, and Morris in the 1850s (Appleby 1992).

It is one of the most remarkable conifers in the Del-aware Valley: it stands 87 feet high (26.5 m) with a spread of 93 feet (28.4 m) and a diameter at breast height of 69 inches (175 cm).

The story of the majestic hardy cedar-of-Lebanon mixes history, geography, plant ecology, horticulture, and a love of conifers. When you next visit the Arnold, Morris, or Tyler Arboreta, take time to enjoy their magnificence and muse on their long journey from the mountains of Turkey to the eastern United States.

Breeding Methods in Crop Plants

Self Pollinated Crops

Mass selection: In mass selection, seeds are collected from (usually a few dozen to a few hundred) desirable appearing individuals in a population, and the next generation is sown from the stock of mixed seed. This procedure, sometimes referred to as phenotypic selection, is based on how each individual looks. Mass selection has been used widely to improve old "land" varieties, varieties that have been passed down from one generation of farmers to the next over long periods.

An alternative approach that has no doubt been practiced for thousands of years is simply to eliminate undesirable types by destroying them in the field. The results are similar whether superior plants are saved or inferior plants are eliminated: seeds of the better plants become the planting stock for the next season.

A modern refinement of mass selection is to harvest the best plants separately and to grow and compare their progenies. The poorer progenies are destroyed and the seeds of the remainder are harvested. It should be noted that selection is now based not solely on the appearance of the parent plants but also on the appearance and performance of their progeny. Progeny selection is usually more effective than phenotypic selection when dealing with quantitative characters of low heritability. It should be noted, however, that progeny testing requires an extra generation; hence gain per cycle of selection must be double that of simple phenotypic selection to achieve the same rate of gain per unit time.

Mass selection, with or without progeny test, is perhaps the simplest and least expensive of plant-breeding procedures. It finds wide use in the breeding of certain forage species, which are not important enough economically to justify more detailed attention.

Pure-line selection: Pure-line selection generally involves three more or less distinct steps: (1) numerous superior appearing plants are selected from a genetically variable population; (2) progenies of the individual plant selections are grown and evaluated by simple observation, frequently over a period of several years; and (3) when selection can no longer be made on the basis of observation alone, extensive trials are undertaken, involving careful measurements to determine whether the remaining selections are superior in yielding ability and other aspects of performance. Any progeny superior to an existing variety is then released as a new "pure-line" variety. Much of the success of this method during the early 1900s depended on the existence of genetically variable land varieties that were waiting to be exploited. They provided a rich source of superior pure-line varieties, some of which are still represented among commercial varieties. In recent years the pure-line method as outlined above has decreased in importance in the breeding of major cultivated species; however, the method is still widely used with the less important species that have not yet been heavily selected.

A variation of the pure-line selection method that dates back centuries is the selection of single-chance variants, mutations or "sports" in the original variety. A very large number of varieties that differ from the original strain in characteristics such as colour, lack of thorns or barbs, dwarfness, and disease resistance have originated in this fashion.

Hybridization

During the 20th century planned hybridization between carefully selected parents has become dominant in the breeding of self-pollinated species. The object of hybridization is to combine desirable genes found in two or more different varieties and to produce pure-breeding progeny superior in many respects to the parental types.

Genes, however, are always in the company of other genes in a collection called a genotype. The plant breeder's problem is largely one of efficiently managing the enormous numbers of genotypes that occur in the generations following hybridization. As an example of the power of hybridization in creating variability, a cross between hypothetical wheat varieties differing by only 21 genes is capable of producing more than 10,000,000,000 different genotypes in the second generation. At spacing normally used by farmers, more than 50,000,000 acres would be required to grow a population large enough to permit every genotype to occur in its expected frequency. While the great majority of these second generation genotypes are hybrid (heterozygous) for one or more traits, it is statistically possible that 2,097,152 different pure-breeding (homozygous) genotypes can occur, each potentially a new pure-line variety. These numbers illustrate the importance of efficient techniques in managing hybrid populations, for which purpose the pedigree procedure is most widely used.

Pedigree breeding starts with the crossing of two genotypes, each of which have one or more desirable characters lacked by the other. If the two original parents do not provide all of the desired characters, a third parent can be included by crossing it to one of the hybrid progeny of the first generation. In the pedigree method superior types are selected in successive generations, and a record is maintained of parent–progeny relationships.

The F2 generation (progeny of the crossing of two F1 individuals) affords the first opportunity for selection in pedigree programs. In this generation the emphasis is on the elimination of individuals carrying undesirable major genes. In the succeeding generations the hybrid condition gives way to pure breeding as a result of natural self-pollination, and families derived from different F2 plants begin to display their unique character. Usually one or two superior plants are selected within each superior family in these generations. By the F5 generation the pure-breeding condition (homozygosity) is extensive, and emphasis shifts almost entirely to selection between families.

The pedigree record is useful in making these eliminations. At this stage each selected family is usually harvested in mass to obtain the larger amounts of seed needed to evaluate families for quantitative characters. This evaluation is usually carried out in plots grown under conditions that simulate commercial planting practice as closely as possible. When the number of families has been reduced to manageable proportions by visual selection, usually by the F7 or F8 generation, precise evaluation for performance and quality begins. The final evaluation of promising strains involves (1) observation, usually in a number of years and locations, to detect weaknesses that may not have appeared previously; (2) precise yield testing; and (3) quality testing. Many plant breeders test for five years at five representative locations before releasing a new variety for commercial production.

The bulk-population method of breeding differs from the pedigree method primarily in the handling of generations following hybridization. The F2 generation is sown at normal commercial planting rates in a large plot. At maturity the crop is harvested in mass, and the seeds are used to establish the next generation in a similar plot. No record of ancestry is kept. During the period of bulk propagation natural selection tends to eliminate plants having poor survival value. Two types of artificial selection also are often applied: (1) destruction of plants that carry undesirable major genes and (2) mass techniques such as harvesting when only part of the seeds are mature to select for early maturing plants or the use of screens to select for increased seed size. Single plant selections are then made and evaluated in the same way as in the pedigree method of breeding. The chief advantage of the bulk population method is that it allows the breeder to handle very large numbers of individuals inexpensively. Often an outstanding variety can be improved by transferring to it some specific desirable character that it lacks. This can be accomplished by first crossing a plant of the superior variety to a plant of the donor variety, which carries the trait in question, and then mating the progeny back to a plant having the genotype of the superior parent. This process is called backcrossing. After five or six backcrosses the progeny will be hybrid for the character being transferred but like the superior parent for all other genes. Selfing the last backcross generation, coupled with selection, will give some progeny pure breeding for the genes being transferred.

The advantages of the backcross method are its rapidity, the small number of plants required, and the predictability of the outcome. A serious disadvantage is that the procedure diminishes the occurrence of chance combinations of genes, which sometimes leads to striking improvements in performance.

Hybrid Varieties

The development of hybrid varieties differs from hybridization. The F1 hybrid of crosses between different genotypes is often much more vigorous than its parents. This hybrid vigour, or heterosis, can be manifested in many ways, including increased rate of growth, greater uniformity, earlier flowering, and increased yield, the last being of greatest importance in agriculture.

Cross Pollinated Crops

The most important methods of breeding cross-pollinated species are (1) mass selection; (2) development of hybrid varieties; and (3) development of synthetic varieties. Since cross-pollinated species are naturally hybrid (heterozygous) for many traits and lose vigour as they become purebred (homozygous), a goal of each of these breeding methods is to preserve or restore heterozygosity.

Mass Selection

Mass selection in cross-pollinated species takes the same form as in self-pollinated species; *i.e.,* a large number of superior appearing plants are selected and harvested in bulk and the seed used to produce the next generation. Mass selection has proved to be very effective in improving qualitative characters, and, applied over many generations, it is also capable of improving quantitative characters, including yield, despite the low heritability of such characters. Mass selection has long been a major method of breeding cross-pollinated species, especially in the economically less important species.

Hybrid Varieties

The outstanding example of the exploitation of hybrid vigour through the use of F1 hybrid varieties has been with corn (maize). The production of a hybrid corn variety involves three steps: (1) the selection of superior plants; (2) selfing for several generations to produce a series of inbred lines, which although different from each other are each pure-breeding and highly uniform; and (3) crossing selected inbred lines. During the inbreeding process the vigour of the lines decreases drastically, usually to less than half that of field-pollinated varieties. Vigour is restored, however, when any two unrelated inbred lines are crossed, and in some cases the F1 hybrids between inbred lines are much superior to open-pollinated varieties. An important consequence of the homozygosity of the inbred lines is that the hybrid between any two inbreds will always be the same. Once the inbreds that give the best hybrids have been identified, any desired amount of hybrid seed can be produced.

Pollination in corn (maize) is by wind, which blows pollen from the tassels to the styles (silks) that protrude from the tops of the ears. Thus controlled cross-pollination on a field scale can be accomplished economically by interplanting two or three rows of the seed parent inbred with one row of the pollinator inbred and detasselling the former before it sheds pollen. In practice most hybrid corn is produced from "double crosses," in which four inbred lines are first crossed in pairs (A × B and C × D) and then the two F1 hybrids are crossed again (A × B) × (C × D). The double-cross procedure has the advantage that the commercial F1 seed is produced on the highly productive single cross A × B rather than on a poor-yielding inbred, thus reducing seed costs. In recent years cytoplasmic male sterility, described earlier, has been used to eliminate detasselling of the seed parent, thus providing further economies in producing hybrid seed.

Much of the hybrid vigour exhibited by F1 hybrid varieties is lost in the next generation. Consequently, seed from hybrid varieties is not used for planting stock but the farmer purchases new seed each year from seed companies.

Perhaps no other development in the biological sciences has had greater impact on increasing the quantity of food supplies available to the world's population than has the development of hybrid corn (maize). Hybrid varieties in other crops, made possible through the use of male sterility, have also been dramatically successful and it seems likely that use of hybrid varieties will continue to expand in the future.

Synthetic Varieties

A synthetic variety is developed by intercrossing a number of genotypes of known superior combining ability—*i.e.*, genotypes that are known to give superior hybrid performance when crossed in all combinations. (By contrast, a variety developed by mass selection is made up of genotypes bulked together without having undergone preliminary testing to determine their performance in hybrid combination.)

Synthetic varieties are known for their hybrid vigour and for their ability to produce usable seed for succeeding seasons. Because of these advantages, synthetic varieties have become increasingly favoured in the growing of many species, such as the forage crops, in which expense prohibits the development or use of hybrid varieties.

Mutation Breeding

Physical Mutagens

Physical mutagens include various types of radiation, viz X-rays, gamma rays, alpha particles, beta particles, fast and thermal (slow) neutrons and ultra violet rays. A brief description of these mutagens is presented below:

Commonly used physical mutagens (radiations), their properties and mode of action.

Type of Radiation	*Main properties*
X – rays	S.I., penetrating and non-particulate
Gamma rays	S.I., very penetrating and Non-particulate
Alpha Particles	D.I., particulate, less penetrating and positively charged.
Beta Rays Particles	S.I., particulate, more penetrating than alpha particles and negatively charged.
Fast and Thermal Neutrons	D.I., particulate, neutral particles, highly penetrating.
6. Ultra Violet Rays	Non-ionizing, low penetrating

Note: particulate refers to particle emitting property DI = Densely ionizing, SI = Sparsely ionizing.

X-rays

X-rays were first discovered by Roentgen in 1895. The wavelengths of X-rays vary from 10-11 to 10-7. They are sparsely ionizing and highly penetrating. They are generated in X-rays machines. X-rays can break chromosomes and produce all types of mutations in nucleotides, viz. addition, deletion, inversion, transposition, transitions and transversions. X-rays were first used by Muller in 1927 for induction of mutations in Drosophila. In plants, Stadler in 1928 first used X-rays for induction of mutations in barley.

Gamma Rays

Gamma rays have shorter wave length than X-rays and are more penetrating than gamma rays. They are generated from radioactive decay of some elements like 14C, 60Co, radium etc. Of these, cobalt 60 is commonly used for the production of Gamma rays. Gamma rays cause chromosomal and gene mutations like X-rays.

Bulk Population Method

Advantages

Bulk method is simple, convenient and inexpensive. 'Natural' selection can be enhanced by growing bulks under stress conditions, e.g., disease, insect, drought. Can use mass selection to modify the bulk. Can replicate bulks and determine performance or test in multiple environments. Can drop bulks at any time.

Disadvantages

Lines selected from the bulk that appear to be segregating may need to be reselected to obtain uniformity. There is a risk that desirable genotypes may be lost from the population; e.g., short plants may be at a disadvantage in competition with tall plants. Competition is not the same as natural selection. Will know little about the genetic structure of the population. Natural selection is not always beneficial.

5

Biological Control

Rice

Biological control as a method of insect pest management in rice began much later than the first success in the control of cottony cushion scale in California in the late 1880s. In fact, information- on natural enemy abundance and diversity, ecology and effectiveness as biocontrol agents, effects of pesticides and cultural practices on their populations, etc.-was very limited at the time. Considerable work has been done during the later half of the 20th century, and the practice of biological control of rice pests is constantly evolving.

Naturally Occurring Biological Control

Rice Insect pests have had long, close associations with their natural enemies, allowing stable relationships to develop. For example, predator and parasitoid guilds recorded in five irrigated rice ecosystems in the Philippines belong to ten orders and 57 families of predators and three orders comprising 40 families of parasitoids, indicating the great diversity and richness of the natural enemy community in rice. The interactions of such predators, parasitoids and insect pathogens are the cornerstone of modern integrated pest management programs in rice. The collective importance of natural enemies to pest population regulation is clearly demonstrated by observing outbreaks of insect pests after the removal of natural enemy communities by a broad-spectrum insecticide.

Conservation

Conservation is the modification of the environment or habitat in combination with judicious use of pesticides in order to conserve natural enemies and enhance biological control. Evaluating the success of biological control is, however, a daunting problem; far different from the "spray-and-count" or "count-and-spray" evaluation methods of pesticide-based approaches. Despite this, environment management to enhance biological control has proven valuable in many settings and appears to be very important for rice insect pest management.

Conservation Biological Control Contains Two Elements:

- Judicious use of insecticides.
- Modification of the environment.

Judicious use of Pesticides

As mentioned earlier, insecticides are more harmful to natural enemies than to pests. There is extensive evidence that the destruction of natural enemies by broad-spectrum insecticides leads to reduced biological control and the subsequent resurgence of some pests. At the early stages of rice crop growth, abundant populations of detritus-feeding and plankton-feeding insects supports large populations of generalist predators in rice fields. The abundance of alternative prey gives the predator populations a "head start" on later developing pest populations. Therefore, insecticide overuse is most damaging early in the season. Insecticide use can also create heavy selection pressure for strains of pests that can even overcome previously resistant cultivars. Such circumstances create outbreaks of secondary pest such as brown planthopper and impair biological control of some key primary pests such as stem borers.

Recent surveys in 10 major rice-growing countries in Asia revealed that pesticide is the dominant pest control tactic of farmers, who use insecticides more frequently than herbicides and fungicides. A large proportion of farmers are still using insecticide compounds classified (by the World Health Organization) as extremely or highly hazardous to human health – namely methyl parathion, monocrotophos, and methamidophos.

The mean number of insecticide sprays per season varies from 0.3 in Laos to 3.9 in Vietnam. Most of these sprays are applied during early vegetative stages mainly targeting leaf feeding insect pests, which usually do not cause economic loss. By and large, insecticides are being misused in rice. Preventive use of insecticides in rice is common and widespread. A common tactic used by farmers to control stem borers is the use of granular insecticides as a preventive method, without any consideration of the pest population density or economic benefit. Farmers' rationales for insecticide use appear to be base on anticipated yield losses due to insect pest attack. Such attitudes are detrimental to biological control, pest management as a whole, the environment and human health. Judicious use of insecticides is essential for natural enemy conservation and for the enhancement of natural biological control.

In order to conserve indigenous natural enemies and enhance natural biological control, the following strategies for insecticide use are recommended:

- Preventive or calendar-based use of insecticides should be stopped.
- Early season spraying during the first 40 days after transplanting should be avoided.
- Selective insecticides should be used only when pest population density reaches a damaging level.
- When using insecticides, use appropriate formulations in the right dose. Apply at the optimum time of intervention following proper application methods.
- Broad-spectrum hazardous insecticides should be avoided.
- If available, use rice varieties that possess host plant resistance to major insect pests.

Modification of Environment

The immediate crop environment can be intentionally modified to favour natural enemies. This requires substantial knowledge of pest and natural enemy ecology, and the effects of cultural practices on them. Habitat management should be aimed at providing environmental requisites to natural enemies such as:

- Alternative hosts or prey.
- Complementary foods – honeydew, pollen, nectar.
- Over-wintering or off-season shelter.
- Modified climate – windbreak.

Some habitat management activities that have a positive influence on the abundance of the natural enemies of rice insect pests are described blow:

Rice Bunds

Vegetated rice bunds support many herbivore species (including a few pests) and many species of natural enemies. They can be a very important source of natural enemy colonization in rice fields just after planting. Rice bunds are important sources of biological control agents of arthropods. Bund vegetation provides alternative and supplementary food for predators and parasitoids. Bunds are especially important for early sown crops and crucial for single-season rice crops. The importance of bund fauna on the early arrival of spiders for biological control of BPH has been demonstrated experimentally.

Bunds also provide shelter for natural enemies at critical times during harvesting, land preparation and between seasons. Puddling (tillage in presence of water) can have drastic effects on natural enemy populations. Many natural enemies take shelter on the rice bunds at puddling and the presence of vegetation on the bunds greatly increases their survival. Manipulation of bund vegetation, particularly the enhancement of broadleaf weeds, can provide vital sources of natural enemies of rice insect pests and can reduce sources of alternate grassy hosts of some diseases such as sheath blight.

Vegetation in Adjacent Habitats

Generally, grasses and other vegetation in habitats adjacent to rice fields serve as habitats of natural enemies and also provide supplementary and complimentary food, over-wintering, or off-season habitats. The precise composition of plant species that will make the greatest contribution to the conservation of natural enemies of rice insect pests has not yet been determined.

Many species of arthropods, including natural enemies, exist concurrently in rice and non-rice habitats. Vegetation in these habitats serves as an important refuge for oligophagous rice parasitoids and generalist arthropod predators.

One such example is Anagrus nilaparvatae (Mymaridae: Heteroptera), an important egg parasitoids of rice planthoppers, which also parasitizes the planthopper Saccharosydne procerus (Delphacidae: Homoptera) of Zizania caduciflora (Gramaineae), which is a favorite vegetable for the residents of the Yangtze Rive Delta in China. Anagrus nilaparvatae can overwinter in S. procerus eggs in Zizania fields during the off-season. Such over-wintering is crucial for the maintenance of natural enemy populations in between the seasons.

Crop mosaics and Staggered Planting

Some non-rice crops such as cowpea, mungbean, maize, bell pepper, garlic, onion, soybean etc. serve as important reservoirs of natural enemies of rice insect pests during the non-rice season. Appropriate crop manipulation in rice growing areas may enhance conservation of natural enemies and thus biological control of rice insect pests. In general, non-rice crop fields forming a mosaic within the rice blocks are good for natural enemy conservation. In a monoculture situation, asynchrony (staggered planting) can provide better conservation of natural enemies than strictly synchronous planting.

Cultural Practices that Disrupt Natural Enemies

Some cultural practices seriously affect natural enemies. These include:

- Burning of rice stubble and/or straw.
- Burning of rice husk on the field prior to onion cultivation (in the Philippines).
- Trimming of non-rice vegetation from rice bunds prior to tillage (puddling).
- Destruction of vegetation on the field during fallow periods.
- Strictly synchronous rice planting over a large area.

From plant nutrient and natural enemy conservation points of view, it is better to incorporate crop residues into the soil rather than burn them. If burning is essential then it should be done in heaps rather than burning residue spread all over the field.

Conservation of Predatory Vertebrates

In many countries insect predatory birds, toads and frogs are caught in rice fields for consumption. Conservation of insect predatory birds, frogs and toads by preventing capture from the rice environment can also enhance biological control of rice insect pests. In Bangladesh, for example, collecting frogs from the lowland rice environment during their breeding period is discouraged.

Augmentation

What is Augmentation?

This method of biological control mainly refers to inundative or inoculative release of mass reared natural enemies. There have been a few attempts, mostly in China, to mass release stem borer egg parasitoids such as Trichogramma japonicum, Telenomus rowani, and Tetrastichus schoenobii. It seems that mass rearing and release is not economical in rice and that this approach has limited prospects for rice insect pest management.

Other Methods of Augmentation

Aside from inoculative and inundative release, a few other augmentation methods have been used with some success. These include the use of perches for insect predatory birds, use of ducks, and rice-fish culture.

Augmentation of Insect Predatory Birds

Placing perches in rice fields can attract some insect predatory birds. One such bird is the black drongo, Dicrurus adsimilis (Bechstein). The black drongo is a purely insectivorous bird and very common in the rice environments of South Asia. Black drongos use perches in rice fields as prey watchtowers and resting sites between prey hunting flights. They hunt any large insects in flight and also prey on insects from the rice canopy. Their prey include, grasshoppers, crickets, moths and butterflies, large caterpillars, etc.. However, they also prey on some beneficial insects such as dragonflies. Black drongos and many other insectivorous birds concentrate in the field for insect predation at tillage, weeding and harvesting stages.

Of about 1140 species of birds known to occur in China, half are entomophagous and about 30 are common and important for pest control in agriculture and forestry. Four species of bird have proven to be effective in grassland locust control in China. These are: the rose-colored sterling Sturnus roseus; the common starling S. vulgaris porphyronotus Sharpe; the chough Pyrrhocorax pyrrhocorax (L.); and the desert wheatear Oenanthe deserti (Temminck). Constructing nest-building sites near target areas can attract these birds.

Use of Ducks

Releasing domestic ducks for pest control is a common component of IPM in China and Vietnam. Ducks are generalist predators, feeding on stem borers, leaffolders, grasshoppers, planthoppers and leafhoppers etc.. Ducks have big appetites and on average one duck can consume more than 100 insects per hour. Ducks are very good at decreasing pest populations quickly, particularly in the early to mid-tillering stage, and also provide partial control of weeds. In China, rearing of ducks for pests control has been found to significantly increase income and reduce pesticide use.

Rice-Fish Culture

Rice and fish association is as old as the development of rice as human food. In the past, rice and free-living fish were very important resources in the lowland environments of Asia. Development of semi-dwarf modern rice plant types has dramatically changed this. In low land environments rice is now mostly grown under irrigation during the dry season. Rice-fish culture in China is almost 2000 years old, but it suffered a setback during the green revolution due to widespread use of toxic insecticides. Increasingly, though, rice-fish culture is regaining popularity to cope with the increasing demand for fish in Asia.

Different types of carp, tilapia and catfish feed on planthoppers and leafhoppers, stem borers (moth and dispersing larvae) or other insects that fall into the water, mosquito larvae and other aquatic insects. Some fish also feed on the outer leaf of the leaf sheath, which contains planthopper and leafhopper eggs. The ecological benefits of rice-fish culture are increased soil fertility, and control of weeds, insect pests and disease. Under rice-fish systems farmers cannot use insecticides, due to potential harmful effects on the fish, which helps to protect natural enemies.

Inundative and Inoculative Release

What is Inundative and Inoculative Release?

A few attempts have been made to control indigenous rice pests through inundative release of parasitoids, with limited success. An attempt to control the dark-headed stem borer Chilo polychrysus in Malaysia by release of Trichogramma japonicum in the early 1930s failed. Such interventions are often unsuccessful due to their high cost and the fact that the in most rice environments insect pests are already being controlled biologically by a rich and diverse group of natural enemies.

Introduction to Classical Biological control

What is Classical Biological Control?

In only two cases did the introduction of exotic parasitoids result in successful biological control in rice production. In the first case, populations of immigrant rice leaffolder, Marasmia exigua (Butler) (Lepidoptera: Pyralidae), in the Fiji Islands were apparently reduced by the introduced larval parasitoid, Trathala flavoorbitalis (Careron) (Hymenoptera: Ichneumonidae) from Hawaii in 1928. The second case involved the control of introduced rice striped stem borers Chilo suppressalis (Walker) (Lepidoptera: Pyralidae) in Hawaii through the introduction of an egg parasitoid – Trichogramma japonicum Ashmead (Hymenoptera: Trichogrammatidae) from Japan and two larval parasitoids – Eriborus sinicus (Holmgren)(Hymenoptera: Eulophidae) and Bracon chinensis Szepligeti (Hymenoptera: Braconidae) from China in 1929. All three parasitoids established and helped to reduce infestation of this stem borer. Several attempts have been made to introduce exotic parasitoids to control indigenous rice insect pests in South and Southeast Asia but none was successful. One such example involved the introduction of a larval parasitoid Paratheresia claripalpis (ven der Wulp) (Diptera: Tachinidae) from Trindad to Malaysia to control the dark-headed stem borer Chilo polychrysus (Meyrick) in 1951.

A more recent example involved the introduction of two egg parasitoids Psix lacunatus (Johnson and Masner) (Hymenoptera: Scelionidae) and Telenomus cyrus (Nixon) (Hymenoptera: Scelionidae) to Palawan Island from the Island of Luzon, Philippines during 1985-87 to control Malayan black bug, Scotinophara coarctata (F.) (Hemiptera: Pentatomitidae). This project was not successful.

Major Natural Enemies of Major Rice Insect Pests

What are the Major Natural Enemies?

The species diversity and total number of natural enemies in tropical rice are impressive, but usually a few species dominate the complex. It is difficult to generalize about the precise role and relative importance of individual species. Here three important groups of insect pests are selected to illustrate the roles that natural enemies can play in the biological control of rice insect pests. The insect pests groups are:

- Stem borers
- Leaffolders
- Leafhoppers and planthoppers.

Rice stem Borers

More than one hundred species of parasitoids, predators and insect pathogens attack rice stem borers. Stem borers are most vulnerable to natural enemies at the egg, neonate larval and adult stages. On average only 1-4% of the egg population reaches adulthood and the rest perish due to biotic and abiotic stresses

Egg Parasitism and Predation

Parasitism and predation of stem borer eggs are usually very high and are important population regulating factors. Three groups of egg parasitoids, Telenomus spp., Tetrastichus spp. and Trichogramma spp., dominate the complex. Some of the major parasitoids are density dependent, meaning parasitism rates increase with increases in egg density, which is an important population regulation mortality factor. The most important egg parasitoids are:

- Tetrastichus schoenobii Ferriere (Hymenoptera: Eulophidae)
- Telenomus rowani (Gahan) (Hymenoptera: Scelionidae)
- Telenomus dignus (Gahan) (Hymenoptera: Scelionidae)
- Trichogramma japonicum Ashmead (Hymenoptera: Trichogrammatidae)Two important egg predators are:
- meadow grasshoppers – Conocephalus longipennis (de Haan) (Orthoptera: Tettigoniidae); and
- crickets, Metioche vittataicollis (Stal) (Orthoptera: Gryllidae).

Larval and Pupal Parasitism and Predation

The complex of larval and pupal parasitoids is large, but not as abundant as that of the egg parasitoids. Stem borer larvae and pupae generally suffer low levels of parasitism.

Larvae and pupae are almost inaccessible to natural enemies because they remain inside the stem. A few parasitoids are equipped with highly specialized long ovipositors that they can use to probe through the rice stem to locate and parasitize stem borer larvae and pupae. The most common larval and pupal parasitoids are:

- Cotesia flavipes Cameron (Hymenoptera: Braconidae)
- Temelucha philippinensis (Ashmead) (Hymenoptera: Ichneumonidae)
- Stenobracon nicevillei (Bingham) (Hymenoptera: Braconidae)
- Bracon chinensis Szepligeti (Hymenoptera: Braconidae)
- Tropobracon schoenobii (Viereck) (Hymenoptera: Braconidae)
- Xanthopimpla stemmator (Thunberg) (Hymenoptera: Ichneumonidae)
- Tetrastichus ayyari Rohwer (Hymenoptera: Eulophidae).

However, young dispersing stem borer larvae suffer high rates of mortality due to predation and desiccation. Disappearance of young larvae is the main reason for stem borer population fluctuations. Important larval predators are:

- Lady beetles, Micraspis spp. (Coleoptera: Coccinellidae)
- Carabid beetles, Ophionea spp. (Coleoptera: Carabidae)
- Rove beetle, Paederus fuscipes Curtis (Coleoptera: Staphylinidae)
- Water bug, Microvelia douglasi atrolineata Bergroth (Hemiptera: Veliidae)
- Water bug, Mesovelia vittigera (Horvath) (Hemiptera: Mesoveliidae)
- Water bug, Limnogonus fossarum (F.) (Hemiptera: Gerridae)
- Ants (Hymenoptera: Formicidae).

Adult Predation

Stem borer moths do not suffer parasitism but may suffer high mortality from a variety of insect, spider and bird predators. Reliable estimates on moth predation in the field situation are not available. The important predators are:

- Anthocorid bug – Euspudaeus sp., (Hemiptera: Anthocoridae)
- Wolf spider – Lycosa pseudoannulata (Boesenberg and Strand) (Araneae: Lycosidae)
- Black drongo – Dicrurus adsimilis (Bechstein)(Dicruridae).

Diseases

Beauveria, Cordyceps and Nomuraea are white fungi that infect stem borers. In high moisture fungal spores germinate and invade the soft tissue and body fluids of hosts and grow out of the body when ready to produce dispersing spores. Important pathogens are:

- Beauveria bassiana (Balsamo) Vuillemin (Moniliales: Moniliaceae)
- Nomuraea rileyi (Farlow) Samson (Moniliales: Moniliaceae)
- Cordyceps sp. (Entomophthoraceae)
- Bacillus thuringiensis Berliner (Bacteria)
- An unidentified nuclear polyhedrosis virus.

Rice Leaffolders

Parasitoids and predators play a very important role in rice leaffolders population regulation. Unlike stem borers, leaffolders are exposed to natural enemies at all the growth stages. Although larva mostly remain inside the leaf fold, entrance of natural enemies into the leaf fold is not very difficult.

Egg Parasitism and Predation

Parasitoids:

- Copidosomopsis nacoleiae (Eady) (Hymenoptera: Encyrtidae)
- Trichogramma japonicum Ashmead (Hymenoptera: Trichogrammatidae)

Predators:

- Cricket – Anaxipha longipennis (Serville) (Orthoptera: Gryllidae)
- Cricket – Metioche vittaticollis (Stal) (Orthoptera: Gryllidae)
- Mirid bug – Cyrtorhinus lividipennis Reuter (Hemiptera: Miridae)
- Lady bird beetle – Micraspis crocera Mulsant (Coleoptera: Coccinellidae).

Larval and Pupal Parasitism and Predation

The larval and pupal parasitoid complex of rice leaffolders is rich. Several dozens of parasitoid species attack larvae and pupae. Common parasitoids are:

- Goniozus nr. trangulifar Kieffer (Hymenoptera: Bathylidae)
- Cotesia angustibasis (Gahan) (Hymenoptera: Braconidae)
- Cardiochiles philippinensis Ashmead (Hymenoptera: Braconidae)
- Macrocentrus philippinensis (Ashmead) (Hymenoptera: Braconidae)
- Trichomma cnaphalocrosis Uchida (Hymenoptera: Ichneumonidae)
- Temelucha philippinensis (Ashmead) (Hymenoptera: Ichneumonidae)
- Xanthopompla flavolineata Cameron (Hymenoptera: Ichneumonidae)
- Tetrastichus ayyari Rohwer (Hymenoptera: Eulophidae).

Reliable information on the extent of larval predation is not available. We do know, however, that the ground beetles Ophionea sp. attack leaf folder larvae.

Adult Predation

Leaffolder moths do not suffer mortality from parasitoids. However, a diverse group of predatory insects, spiders and vertebrates predate on rice leaffolder moths. The extent of moth predation in nature is not known.

Disease

Entomopahogenic bacteria such as Bacillus thuringiensis Berliner and a granulosis virus cause mortality of rice leaffolder larvae.

Rice Leafhoppers and Planthoppers

In general the impact of predators on rice leafhopper and planthopper populations is stronger than that of parasitoids.

Egg Parasitism and Predation

Egg parasitism varies considerably among hopper species and fluctuates during the season. Leafhopper and planthopper eggs are generally more heavily parasitized than are adults and nymphs. The major egg parasitoids are:

- Oligosita yasumatsui Viggiani et Subba Rao (Hymenoptera: Trichogrammatidae).
- Anagrus spp. (Hymenoptera: Mymaridae).
- Gonatocerus spp. (Hymenoptera: Mymaridae).

The most abundant predator is the green mired bug Cyrtorhinus lividipennis Reuter (Hemiptera: Miridae), which predate on both eggs and nymphs. C. lividipennis appears to play an important role in regulation of plant and leafhopper populations.

Nymph and Adult Parasitism and Predation

The major parasitoids of planthoppers and leafhoppers are dryinid and strepsipteran wasps. Parasitism varies between 10% and 50% depending on the season and location. Important parasitoids are:

- Pseudogonatopus spp., (Hymenoptera: Dryinidae)
- Haplogonatopus apicalis Perkins (Hymenoptera: Dryinidae)
- Echthorodelphax fairchildi Perkins (Hymenoptera: Dryinidae)
- Elenchus yasumatsui Kefune and Hirashima (Hymenoptera: Strepsiptera)
- Tomosvaryella spp. (Hymenoptera: Pipunculidae)
- Pipunculus spp. (Hymenoptera: Pipunculidae).

A complex of predators predate on nymphs and adults of rice leafhoppers and planthoppers. Among them spiders and predatory insects are important.

The important predators are:

- Wolf spider – Lycosa pseudoannulata (Boesenberg and Strand) (Araneae: Lycosidae)

- Water bug – Microvelia douhlasi atrolineata Bergroth (Hemiptera: Veliidae)
- Lynx spider – Oxyopes spp. (Araneae: Oxyopidae)
- Long-jawed spider – Tetragnatha spp., (Araneae: Tetragnathidae)
- Rove beetle – Paederus fuscipes Curtis (Coleoptera: Staphylinidae)
- Ground beetle – Ophionea nigrofasciata (Schmidt-Goebel) (Coleoptera: Carabidae).

Diseases

Several fungal pathogens infect rice leafhoppers and planthoppers. Prolonged high moisture is necessary for the germination of fungal spores. The fungus grows into the body of the infected hopper; when the insect is dead, fungus grows out of the body through the joints.

One mermithid nematode also infects hoppers and after death of the insect the nematode emerges from the body. The incidence of nematode parasites is usually low, but relatively higher in the wet season than the dry season. Important pathogens are:

Fungus

- Metarhizium anisopliae (Metchnikoff) Sorokin (Moniliales: Moniliaceae)
- Metarhizium flavoviride Gams and Roszypal (Moniliales: Moniliaceae)
- Beauveria bassiana (Balsamo) Vuillemin (Moniliales: Moniliaceae)
- Hirsutella citriformis Speare (Moniliales: Stilbaceae)
- Erynia delphacis (Hori.) (Entomophthoraceae)
- Entomophthora coronata (Constantin) Kevorkian (Entomophthoraceae).

Nematode

- Hexamermis sp. (Mermithidae)
- Agamermis unka Kaburaki & Imamura.

Integrated Pest Management

Growing pest resistant rice varieties-Gallmidge and BPH resistant varieties with other desirable traits are available now. Depending on the region (key pest of the area) variety should be opted.

- Nursery protection by applying carbofuran granules @ 40 kg/ha or Phorate @ 12.5 kg/ha of nursery bed before 5-7 days of pulling the seedlings in endemic areas of gallmidge and stem borer only.
- Transplanting at appropriate stage after removal of terminal part (2 to 4") of seedlings to reduce the chances of carrying and migration of immature stages of yellow stem borer and leaf folders. Leaving alleys after every 5 metres in the main field is advantageous for field monitoring as well as manual movement to take up required interventions.
- Monitoring and fixed area (bench mark site) surveys to understand pest and associated beneficial fauna development. Adoption of prescribed sequential sampling plans for assessing different stages of plant and leaf hoppers as well as stem and leaf damages is ideal.
- Using pheromone traps (8 per acre) and light traps to understand the brood initiation of yellow stem borer, leaf folder and climbing cut worm should form a basic step in each situation.
- Adoption of mechanical practices like rope running to expose case worm and leaf folder larvae before going to chemical interventions.
- Regulating 'N' applications to avoid nutritional resurgence of plant and leaf hoppers. Supplementing N with K2O and neem coated materials for better utility and reduction of negative effects.
- Application of recommended insecticides-Endosulfan/Monocrotophos/Carbaryl/Chlorpyriphos/Quinalphos to reduce leaf folders and climbing cut worm as well as yellow borer if situation warrants.
- Harvesting of crop to ground level to reduce the chances of yellow stem borer and gallmidge build up, since stubbles serve as reservoir of these pests under many situations.
- Creating avenues for broad spectrum predators like spiders by maintaining shelters in and around rice fields during off season. Similarly rice crop maintenance should be in the process of conserving and exploiting of potential predators like mirids, dragon flies and carabid beetles, common in rice eco-system.

Management of Diseases

Sheath Blight

Use of disease free seed. Avoid close spacing, seed treatment with Carbendazim 50 WP @2.5 g/kg seed or soaking seed in Carbendazim @1g/l of water for 24 hours or spraying Propiconazole 25 EC (Tilt) @ 1ml/l or Hexaconazole 5 EC (Cantof) @2ml/l or Validamycin 3@ 2ml/l.

Bacterial Blight

Use of disease free seed. Grow tolerant varieties like Swarna and MTU 4870. No chemical is effective. Avoid excess N and water.

Tungro Virus

Cultivate tolerant varieties like MTU 9992, IET 7301, IET 7302, IET 8558, MTU 1002 to MTU 1005. Removal and destruction of affected plants. Vectors (GLH) can be controlled with Carbofuran 3G (10 kg/acre)/Monocrotophos 2ml/l or Chlorpyriphos 2 ml/l.

Blast

Cultivate tolerant varieties like IR 36, IR 64, MTU 9991, MTU 9992, MTU 7014, NLR 28600, Pinakini, Tikkana etc. Treat the seed with Carbendazim 50 WP @ 2.5g/kg seed or with Tricyclozole @ 1g/kg seed. Spray Ediphenphos 50 EC @1ml/l or Trycyclozole 75WP @0.6 g/l soon after observing symptoms and another spray 15 days after.

Wheat

Wheat is the most important food grain crop in India, occupying approximately 230 million ha world over. India is the second largest producer of the wheat with 25 million ha area and 70 million tones production. About one tenth of the global production is from India. Wheat is grown during winter in the states of Punjab, Haryana, Uttar Pradesh, Bihar, Rajasthan, Madhya Pradesh and Maharashtra. While in northern states it is grown as irrigated crop, in parts of Madhya Pradesh and Maharashtra it is grown largely under rainfed conditions with little or no irrigation support. In Maharashtra wheat is grown in 7 lakh ha with average productivity of 13.2 q/ha against the national average of about 26.5 q/ha. Out of the 4 species, *Triticum aestivum* occupies almost 90% of the total wheat area.

Soil Requirement

Well drained loams and clay loams are ideal for wheat cultivation. However, good crops are raised in sandy loams and black soils also. Soil pH below 6.5 and above 7.8 are not suitable.

Season and Duration

Being low temperature loving crop, it is grown in winter from end of October to February under rainfed conditions and from middle of November to March/April under irrigated conditions. Total duration of the crop ranges from 110 to 120 days under irrigated conditions and 100-110 days under rainfed conditions.

The major diseases of wheat in India are, three rusts-leaf, yellow and stem rust, Karnal bunt, foliar blights, powdery mildew and loose smut. Diseases of limited importance include head scab, foot rot and flag smut; these diseases though of lesser importance, may be important in certain pockets.

Puccinia Recondita Tritici

Distribution: Throughout wheat growing regions of India.

Development: Pathogen over-summers in low and mid altitudes of Himalayas and Nilgiris. Primary infections develop from Wind deposited urediospores in eastern Indo-gangetic plains in middle of January where it multiplies and moves westwards by March. Temperatures of 20:t 5° C with free moisture (rain or dew) cause epidemics. Severe infection causes upto 30 percent yield losses.

Management: The presently recommended varieties in most of the wheat growing zones are rust resistant.

Puccinia Striiformis Tritici

Distribution: Hills, foothills and plains of north western India and southern hills zone (Nilgiri hills of Tamilnadu).

Development: Spreads through air-borne urediospores, when temperature are 10-20°C but the spread is checked above 25°c. Pathogen survives in the cool temperatures of hills (Himalayas and Nilgiris) and the primary infection takes places by middle of January in the foot hills and sub mountainous parts of north western India.

Also, infection comes from across the western border, hence the probability of evolution of new races increases in this area. Yellow rust from Nilgiri hills cannot come out of the zone due to high temperatures in the Peninsular and Central India.

Management: Most of the presently recommended varieties are resistant. Major emphasis is on host resistance and cultivation of resistant varieties is the main strategy of management.

Stem Rust/Black Rust

Puccinia Graminis Tritici

Distribution: Mainly in Peninsular and I Central India, may occur in traces in Northern India too' were the infestation comes late.

Development: Develops from air-borne urediospores, needs free moisture and temperature above 20° C for spread. It can cause severe grain losses if infection is early. The pathogen perpetuates in Nilgiri hills during off season and becomes air-borne. If Peninsular and Central India experience rainfall during November then epidemics are severe. Late infections cause less damage in north India.

Management: The presently re-commended varieties in most of the wheat growing zones are rust resistant, hence the old susceptible varieties be avoided.

Karnal Bunt

Tilletia Indica (=Neovossia Indica)

Distribution: Parts of Northern Plains, especially Punjab, parts of northern Haryana, foot hills of J&K and HP., tarai area of Uttranchal, in lesser severity in Rajasthan, Bihar and UP. The states of Gujarat, Maharashtra, Karnataka and several parts of M.P. are free of KB.

Development: Seed and soil-borne; infection occurs at flowering by means of soil-borne inoculum. The degree of disease development depends upon the weather conditions prevailing during spike emergence to grain filling stage of crop. If the rains occur during the month of February in north Indian plains (disease-prone areas), the disease is likely to come with higher severity.

Management: Among the present day varieties, PBW 502 is resistant while the others show various levels of susceptibility. For management of this disease, one spray of Propiconazole (Tilt 25EC@ 0.1 %) should be given at the time of anthesis. Integration of one spray of propiconazole with one spray of bioagent fungus, *Trichoderma viride* (0.4% suspension) gives almost cent per cent disease control. The bioagent spray should be done before earhead emergence (Crop growth stage 31-39 on Zadoks scale), followed by the spray of chemical at start of earhead emergence (crop growth stage 41-49 on Zadoks scale). Two sprays of *T. viride,* at these two critical growth stages also give non chemical control of the disease which is almost similar to one spray of propiconazole. Chemical control should be adopted mostly in seed production plots.

Black Point

Alternaria Alternate

Development: Disease causes blackening of embryonic region of the seed (black point), discoloration of area beyond the embryonic region (black discoloration *(Caused by Aalternata, Curvularia lunate, Epicoccum sp., Bipolaris sorokiniana, etc.)* and eye-spot symptom (B. *sorokiniana).* The warm and humid weather at grain filling or near maturity favours this disease.

Management: This disease is of minor importance. Only when the disease percentage is high, it causes concern to the trader and the consumer. The discolored seeds are mostly shrivelled and they are separated out during processing.

Loose Smut

Ustilago Segatum (U. Tritici)

Distribution: North Indian plains and northern hills zone.

Development: It is a seed borne disease; infection occurs during *Loose Smut* flowering through wind-borne spores. The infection remains dormant inside the otherwise healthy looking seed but the plants grown from such seeds bear infected inflorescence. Infection is favoured by cool, humid conditions during flowering period of the host plant.

Management: Disease can be easily controlled through seed treatment with systemic fungicides hence resistance breeding has not attracted much attention. Treat the seed with fungicides like carboxin (Vitavax 75WP @ 2.5g/kg seed), carbendazim (Bavistin 50WP @ 2.5g/kg seed), tebuconazole (Raxil 2DS @ 1.25g/kg seed) if the disease level in the seed lot is high. If it is low to moderate, treat the seed with a combination of *Trichoderma viride* (@4 g/kg seed) and half the recommended dose of carboxin (Vitavax 75WP @ 1.25g/kg seed).

Foliar Blights

Distribution: Mainly in eastern India but also occurs in Peninsular and Central *Foliar blights* India. This disease complex is emerging as a problem in the north western India too.

Development: The disease requires high temperature and high humidity. This disease is more severe in late sown crop and causes substantial yield losses through formation of shrivelled grains. Most of the varieties are susceptible or moderately susceptible. The disease can be controlled through one spray of propiconazole (Tilt 25EC @ 0.1 %).

Powdery Mildew

Erysiphe Graminis Tritici

Distribution: Mainly in the cooler areas and hilly region; foot hills and plains of north-western India and the southern hills (Nilgiris).

Development: Powdery mildew can easily be diagnosed by the white, powdery patches that form on the upper surface of leaves and stem. With age, the patches turn dull dirty white and may have small black specks embedded.

This disease can spread to all aboveground parts of the plant, including earhead and awns. The disease infects plants during periods of high humidity (not necessarily rain) and cool to moderate temperatures. Low light intensity, which accompanies dry weather and a dense crop canopy favours this disease.

Management: Present day varieties are not resistant to powdery mildew. Hence, the disease severity is more in some pockets. Avoid excessively dense, stands by using adequate seed. For chemical control one spray of propi-conazole (Tilt 25EC@ 0.1 %) on disease appearance (which usually occurs during early March in northern plains) is highly effective.

Head Scab

Fusarium Graminearum

Distribution: Parts of Punjab, especially in the sub mountainous regions. Bread wheat suffers lesser damage than the durum. It was first recorded in severe proportion in some parts of Punjab during 1995-96 crop season and again during 2004-05 crop season.

Development: Disease development is favoured by cool, moist weather with high humidity.

Spores are produced on crop debris and reach the leaves through rain splash or wind. Apart from ear head infection, it can cause seedling blight and foot rot leading to lodging.

In severe cases, it can cause shriveling of grains and low-test weights. At present, it is a disease of limited importance but has the potential to emerge as a major problem due to the production of toxins.

Management: Bread wheat are more resistant than durum. However, no resistant varieties are available. Hence, vigil is needed for this disease.

Cereal Cyst Nematode

Heterodera Avenae

Distribution: This nematode is found In most of the cereal growing regions of the country, especially, the dry and warmer areas of Rajasthan, Haryana and Punjab, but incidence is less in cooler climates.

Development: Larvae enter the roots near the growing point especially, at the seedling stage. Roots of infected plants become predisposed to various soil borne disease like root rots. In infected fields, the losses can be considerably enough.

Management: Most of the wheat cultivars are susceptible but some resist cyst formation. Chemical pesticides, some natural plant products and botanicals, coupled with improved cultural practices help in management of the CCN. For Rajasthan, one CCN resistant variety, CCNRV-1 is available for the disease prone areas.

Seed Gall Nematode/*Ear Cockle*

Anguina Tritici

Distribution: It is found mainly in some parts of northern India especially the states of Bihar, Jharkhand, eastern UP and Chhatisgarh.

Development: These nematodes are spread through seed galls in the seed lots during planting and harvesting. Wet weather favours larval movement and infestation. The nematode invades the crown and basal stem area, finally penetrating floral primordia. This leads to formation of nematode galls in ear heads.

Management: Use of clean seed (free of galls) is the only method to prevent this disease. For removal of galls, the seed lots are floated in 2-5 per cent brine solution. The galls, which float on the surface, can be easily separated and destroyed away from the fields. The seed thus cleaned should be washed with fresh water and used for planting.

Aphids

Sitobion Avenae, Rhopalosiphum Padi and Various other Species

Development: The aphids exist in different stages, *viz.,* winged (alates), wingless (apterous) sexual and asexual forms. The rapid spread takes place through asexual reproduction where females give rise directly to nymphs rather than eggs. Infestation usually occurs during second fortnight of January till crop maturity.

Management: When feeding in sufficient numbers, they can cause considerable damage, but under normal conditions, losses are not much. Chemical pesticides are recommended for this pest in wheat if the level of aphids per tiller crosses 10 during vegetative phase and 5 during reproductive phase.

However, there is need to keep watch on this pest. The spray of imidacloprid @ 20 g a.i. per ha initially on border rows and if infestation is severe then in entire field will give good protection against this pest. Generally, natural enemies present in the field help in controlling the population of this pest.

Brown Wheat Mite

Petrobia Lateens

Distribution: In most of the wheat growing areas, under rainfed conditions, especially in the states of Rajasthan, Haryana and M.P. Sometimes, it is a pest in humid and warm conditions of irrigated areas also.

Development: They *Brown wheat* cause damage through *mite infestation* sucking mouth parts. When present in large numbers, mites cause a silvery flecking on leaves. Individual mites are too small to be visible with naked eye without 'effort. These can be seen by shaking the infested leaves on a white paper.

Management: Most of the times, mites do not cause any production constraint in wheat so no management practices are required. However, there is a need to keep vigil on this pest so that it may not become important in changing cropping sequence of future.

Army Worm

Mythimna Separate

Distribution: Mostly in the warmer climates of central India and to some extent in northern plains.

Development & Management: The larvae are found in the cracks of soil and hide during the day but feed during night or early morning. In wet and humid weather, they may feed during day time also. They survive during summer on the subsequent crops like rice and also continue to exist in rice stubbles before wheat crop comes in the field. Recently, this pest is catching attention in the northern India under Rice-Wheat rotation and where rice stubbles/straw remain in the fields.

Legume Pod-borer

Helicoverpa Armigera (= Heliothis Armigera)

Distribution and Importance: This is a polyphagous insect that attacks various legumes as a pod border.

It is seen damaging wheat ear heads at grain development stage when major hosts are not available. However, the damage is below economic threshold level. It is found mostly in northern and central parts of India. Wheat can serve as a bridge host for carry over of this polyphagous pest.

Termites

Odontotermis Obesus, Microtermis Obesi

Distribution: Mainly in the northern and central India, but also in some pockets of peninsular India.

Early Symptoms of Damage: Termites attack the crop at various growth stages, from seedlings to maturity. The severely damaged plants can be easily uprooted and look wilted and dried. In case roots are partially damaged, the plants show yellowing.

Management: For effective management, chemicals like endosulfan, chlorpyriphos and carbosulfan can be used both for seed treatment and for broadcast of treated soil in standing crop.

Important Pests and Diseases

Important insects pests of wheat are termite, Gujhia weevil, cutworms, armyworm, thrips, aphids, shoot fly and stem borer. Among diseases, black, yellow and brown rust, loose smut, kernel bunt, leaf blight and powdery mildew are common.

Pest Management

Habitat Management and Cultural Practices – Biodiversity is a key component in keeping the pest's population below ETL. Intercropping of mustard and Rajgira and crop rotation with Bajra-wheat-Moong-Jowar-wheat or legumes-wheat-Moong-Bajra/Jowar + legumes-wheat can effectively check the spread of insect pests.

Jaggery (10 kg/ha) applied with 500 lit Jeevamrut attract ants that predate on aphids, thrips and larva. Random planting of maize or installation of about 10-12 bird perches in one ha attract birds that predate on the pests.

Insect-Pest Management

Termites: Termites can damage the crop at any stage of crop growth. Problem is more predominant in rainfed areas than irrigated ones. Use of undecomposed FYM under un-irrigated conditions can also increase the chances of termite attack. Use of Neem leaf manure (5 q/ha) or Neem seed manure (1 q/ha) in the soil at the time of sowing can prevent termite attack.

Gujhia Weevil: Gujhia weevil attack germinating seedlings, resulting into high seedling mortality. Application Neem leaf/seed manure can prevent their attack.

Army Worms: Caterpillars of this insect feed on plants particularly tender parts during the night and hide during the day time. They also damage leaves and ear-head. Spraying Neem leaf extract (5 kg Neem leaves crushed in boiled water and diluted to 100 liter) can effectively reduce the attack.

Brown Wheat Mite, Aphids and Jassids

Intercropping of wheat with mustard and safflower (100 gm seed of each mixed with 100kg wheat at the time of sowing) can effectively check the spread of mites. In case of severe attack, ferment 15 lit cow urine, 2 kg cow dung and 15 kg crushed Neem leaves in 100 lit water for 3-5 days. Filter the fermentate and use as foliar spray over one acre. This formulation is safe for natural enemies of pets such as lady bird beetle etc.

Rodent management – Rodents can be controlled by inserting a thorny twig of *Acacia* up to 1.5 to 2feet in each rat hole over the entire farm can prevent the rat menace. Bird perches (about 10-12/ha) attract owls at night that feed on rats. Use of horse gram flour balls mixed with cotton thread or flour mixed with cement powder, is very common. Rats readily eat these balls, resulting into blocking of their anus, which results into death. Rats that cannot excrete also get irritated and bite other rats in the hole, forcing them to run away.

Disease Management

Wheat crop grown organically is generally free from diseases. But diseases that affect conventional farms can appear occasionally during the conversion to organic.

Rusts-Rusts are caused by three different species of the fungi, Puccinia. Brown and yellow rusts are particularly important in the north-western India. Black rust appears quite late in these areas and normally do not cause substantial damage except in wheat fields sown very late. However, black rust appears in severe form in southern, central and eastern India and cause significant damage.

1. *Brown-Rust:* The small, round-oval spots (called pustules) of brown rust (caused by *Puccinia recondita tritici*) develop mainly on the leaves but under severe attack, appear on sheath and stem as well. Bright orange spots appear either in small clusters or irregularly scattered spreads very fast on susceptible varieties during warm and humid environmental conditions. In case of severe attack, plants mature early with poor root development and produce light and shriveled grains. This rust is most widespread in the country and generally the most damaging in many areas.
2. *Yellow Rust*: This rust is caused by the fungus, *Puccinia strciformis*. Yellow spots develop mainly on the leaves, but in case of severe attack, spots can also appear on leaf sheaths, stalks and glumes. The spots are oval in shape and lemon-yellow in colour and are formed in long streaks or rows, and smaller than those of brown rust. Appearance of lemon-yellow pustules in rows is characteristic feature of yellow rust. Under severe attack, these rows are not distinct, and large part of the plant gets covered with crowded pustules. Yellow rust is predominant in hills, foot-hills and north-western part of the country where the temperatures are low. In plains it does not cause much damage.
3. *Black Rust*: Black rust of wheat is caused by *Puccinia graminis tritici*. The spots occur on stem, leaf sheath, leaves and ear-heads but the stem is often most severely affected. The spots are reddish-brown in colour and elongated in shape. It spreads fast under relatively warmer and humid climate. As the crop matures, dark, black, elongated spots are formed. Black rust, though prevalent all over the country, appears in epidemic forms in the southern, central and eastern India, which are characterized by high temperatures in the wheat growing season.

Management of Rusts

Most effective method for control of rust, is to grow rust-resistant varieties. Biodiversity among wheat cultivars can also effectively check the rust problem. Use 3-4 wheat varieties at a time on each farm. Avoid late sowing or late maturing varieties. To protect the crop from rust infection, spray 5 liters of sour buttermilk mixed with 200 liter of water. Amaranth (chaulai or lal bhaji – a common green leaf vegetable) or Mentha (Pudina) leaf dust can also be used as fine spray (25-30 gm dry leaf powder per lit of water) to prevent the infection of rusts. Foliar spray of dry leaf extract of *Hibiscus rosa-sinensis* (China rose) can also prevent the rust infection.

Loose Smut:-It is caused by a fungus, *Ustiloto tritici.* Generally, diseased plants are indistinguishable from the healthy plants before heading, but diseased plans produce heads earlier than healthy plants. However, diseased plants of wheat variety Sonalika (RR-21) can be detected before ear emergence also by a characteristic streaking and yellowing of the flag leaf which starts drying from the tip and later becomes brownish in colour. Flag leaves of diseased plants of Sonalika can easily be seen in the field from a distance.

Terminal symptom in all the varieties is the production of black powder in place of wheat grains in the almost every ear of the diseased plant. The black powder is blown to flowers of healthy plants, they infect the mature grain and remains dormant until germination. Infected grains cannot be distinguished superficially from the healthy ones. After wowing these grains, the fungus gets activated and grows inside the seed first behind the growing point, keeping pace with apex of the growing shoot. As the ear formation starts, there is an accumulation of the fungus in the floral parts, which get completely destroyed due to formation of the black powder. Use of resistant varieties is the best option under organic farming. Also as the disease is seed borne the use of disease free seed can prevent its occurrence. In case of doubt, treat the seed with 5%Vermiwash. For seed production, visit the field daily for early symptoms of disease. Uproot infected plants and burn them before their spores are dispersed. Solar heat treatment of seeds in hot summer areas of plains can kill the pathogen to a great extent. On a bright sunny day in the last week of May or first week of June, soak the seeds in water for about 4 hours in the morning and then dry it under hot sun in the afternoon. Store the well-dried seeds for use in next season. The seed soaking activates the dormant fungus in the seed and heat of the sun kills it.

Bunt:-Kernal bunt is a common problem in northern India, in all the commercial varieties of wheat, but the disease is rare in traditional ones. The disease reduces both quantity and quality of wheat. Flour prepared from wheat with more than 3 % infected grains smells fishy and is unsafe for consumption. This disease is caused by fungus *Neovossia indica*. The disease can be recognized only after the grains have developed. A portion of the grain along its groove is converted into a black powdery mass, which gives a foul smell. In a plant hill, all the ears are not affected and in an ear only few grains are infected.

Growing resistant varieties is best option. Other precautionary and control measures are as follows:.

(1) Avoid planting highly susceptible varieties like HD-2009 (Arjun), UP-319, WL-711 and HD-1982 (Janak) in the Tarai areas-Himalayan foothills.

(2) Use disease free seeds. Pre-treat seeds with 5%Vermi-wash

(3) Use of Mustard flour 1 kg and milk 5 liters mixed with 100 liter water as foliar spray.

(4) Avoid continuous cropping of wheat in the same field. Practice field sanitation and deep ploughing in summer.

(5) Avoid excessive irrigation particularly during ear formation.

Harvesting

Harvesting is to be done when the ears are sufficiently dry and grains have approximately 15% moisture. Except some parts of Punjab and Haryana, wheat is harvested manually and threshed by wheat threshers. In some area bullocks are also used for threshing.

Sometimes cutter wheels are also used for crushing wheat plants and separating wheat from chaff. In some parts of Punjab and Haryana combined harvester does all these operations in one go. But in this process recovery of grain is less and the straw is spread over the field.

Maize

- Pest is an organism that causes considerable damage to plant and draws the attention of man.
- Over 130 insect pests have been reported to infest maize but only very few are serious and require management. Amongst these, the most serious pests are the stem borers.
- The spotted stem borer, *Chilo partellus* is the key pest throughout India and the pink borer, *Sesamia inferens* is serious pest in Peninsular India.
- Shoot fly species *Atherigona soccata* and *A. naqvii* found to be serious in spring maize in Northern India.

- The losses ranging 26.7 to 80.4 percent have been reported due to *C. partellus*. Losses due to *S. inferens* found varied from 25.7 to 78.9 percent.

Nature of Damage

- Maize plants at the end of mid whorl stage are usually attacked.
- Aphids suck sap from plants and cause yellowing and mottling. Diseased plants may become stunted and turn reddish as they mature.
- If young plants infected they seldom produce ears. The aphid colony may some time cover completely the emerging tassels and the surrounding leaves preventing the emergence.
- Ears and shoots are also infested and seed set may be affected. Honey dew excreted by aphids favours the development of sooty molds.

Control

- One spray of systemic insecticides like dimethioate or monocrotophos at 0.1% will controls the aphids.

Pink Borer: *Sesamia Inferens*

Occurrence

- A serious pest in peninsular India more severe in rabi.
- Other hosts include Ragi, Sorghum, Bajra, Wheat, Paddy and Sugarcane.
- Moth is medium sized stoutly built having straw coloured forewings with a mid longitudinal dark brown broad triangular streak. Hind wings are white.

Life Cycle

- Round pearl like yellowish eggs ranging 80-300 are layed in 2-3 longitudinal rows usually with in the sheaths of bottom leaves of young maize plant of two to three weeks old. As the time for hatching approaches, eggs become brown or shy grey.
- Full grown larvae are stout smooth about 25 to 30 mm in length purplish pink on the dorsal side and white on ventral side. Life cycle is completed in 6-7 weeks with 4-5 generations in a year.

Nature of Damage

- Newly hatched larvae remain in group behind the leaf sheath and begin chewing on the stem and epidermal layer of the sheath. Some migrate to neighboring leaf sheaths, while others penetrate the stem.
- Whorl feeding of larvae results in rows of oblong holes in unfolding leaves unlike round shot holes produced by *Chilo partellus.*

  o ***Cob and Tassel at oneplace*:** Later they bore in to central shoot resulting in the drying up of the growing point and formation of dead heart in young maize plant as a result of larval feeding some times the bottom internodes show circular ring like cuts.

  o Severe damage causes the stem to break. Severely infected plants due to stunting may appear to have some times the cob and tassel at one place.

Management of Stem Borers

- Pest management strategies viz., cultural, biological, host plant resistance, chemical and integrated pest management for maize stem borers can be practiced to effectively control both stem borers *Chilo Partellus* and *Sesamia inferens.*

Farm Sanitation

- Collection and burrying stubble and stalks or ploughing and destroying crop residue, removal of infested plant parts or infested plants at hoeing are recommended.

Inter Cropping

- Growing maize in association with various legumes significantly reduced *C. partellus* damage in maize. Also intercropping maize with soyabean gave a considerable reduction in *C. partellus* and *S. inferens* and 9.2% increase in grain weight per plant.
- Therefore intercropping has a great potential in reducing the pest incidence.

Exploitation of Ovipositional Behaviour

- The ovipositional behaviour of *C. partellus* could be exploited for its management.
- Clipping of lower leaves of maize (up to fourth) on which most of eggs of *C. partellus* are laid, reduced damage to some extent.

- As this practice may not be practicable in a large scale, they recommended earthing-up of the crop to bury the lower leaves as a possibility for reducing initial infestation.

Plant Resistance

- Intensive screening to identify resistance sources against (*C. partellus* yielded an important germplasm source. Antigua Gr. 1 and few breeding lines.
- Sources of tolerance to *S. inferens* have been identified. These materials are currently being used in Directorate of maize Research and also in other centres breeding programs.
- At this time, however, stem borer resistant cultivars/hybrids have yet to be released in the semi arid tropics.
- Further more, the work on the screening for resistance is being done extensively and continuously and there is every possibility of identifying sources of resistance to borers and development of resistant hybrids/varieties.

Biological Control

- Many natural enemies have been reported to attack *C. partelllus*.
- Among the parasitoids reported, *Trichogramma* spp. as egg parasitoids and *Aphanteles favipes* and *T. Chilonis* and larval parasitoids are important.
- *T. cllilonis* has been found parasitising egg masses of *C. partelllus* upto 70%. *A. flavipes* is the most dominant and most widely distributed larval parasitoid of *C. partellus* in India with a parasitisation ranging from 5.3-42.8%.
- Even though a number of natural enemies have been reported on maize borers, but in general, the efficiency of natural enemies in particular farming environments is not known.
- The scope for successfully controlling maize stem borers with natural enemies is limited by the short cropping period and the lack of continuous habitat for natural enemies.

Chemical Control

- Seed treatment
  Seed treatment with Carbofuran 35 ST dust.
- Seed treatment with Carbofuran (40F)@ 2.5g/kg of seed was also found effective against *S. inferens*.

Effective Control Schedule

- 1st Foliar application of Endosulfan 0.1% at 15 days after sowing
- 2nd application of 4% granules al 15 kg/ha a fortnight later against *C. partellus*.

Stem Borer: Chilo Partellus

It is the most destructive pest of maize and Sorghum in Srilanka, India, Pakistan, Afghanistan, Uganda, Central and East Africa.Pest is found in throughout India.

Nature of Damage

- The young larvae first feed on the leaves, making a few shot holes and then bore their way downwards through the central whorl as it opens.

Shot Holes on Leaf

- More shot holes become visible, indicating an earlier attack and the plant also shows dead-hearts. Caterpillars damage maize by boring into the stems, cobs or ears.
- Stem borer infestation results in pinhole damage in the leaves initially and "dead hearts" subsequently.

Life-cycle

- The insect breeds actively from March-April to October and for the rest of the year it remain in hibernation as a full-grown larva in maize and sorghum stubble, stalks or unshelled cobs.
- The larvae pupate sometime in March and emerge as moths in the end of that month or in early April.
- They are active at night, when they mate and lay eggs on the underside of the leaves of various host plants, particularly the early crops of maize and sorghum grown as fodder.
- The eggs are flat, oval, and yellowish and are laid in overlapping clusters each containing up to 20 eggs. A female lays over 300 eggs during its life-p of 2-12 days, and the eggs hatch in 4-5 days in summer.
- The larva become full-fed in 14-28 days passing through six stages and after making a hole in the stem pupates inside it.
- The life cycle is completed in about 3 weeks and there are probably 5 generations in a year. The full-grown caterpillars

of the last generation hibernate in stubble, stalks, etc, and remain there till the next spring.

Management of Stem Borers

- Pest management strategies viz., cultural, biological, host plant resistance, chemical and integrated pest management for maize stem borers can be practiced to effectively control both stem borers *Chilo Partellus* and *Sesamia inferens.*

Cultural Control

Farm Sanitation

- Collection and burrying stubble and stalks or ploughing and destroying crop residue, removal of infested plant parts or infested plants at hoeing are recommended.

Inter Cropping

- Growing maize in association with various legumes significantly reduced *C. partellus* damage in maize. Also intercropping maize with soybean gave a considerable reduction in *C. partellus* and *S. inferens* and 9.2% increase in grain weight per plant. Therefore intercropping has a great potential in reducing the pest incidence.

Exploitation of Ovipositional Behaviour

- The ovipositional behaviour of *C. partellus* could be exploited for its management.
- Clipping of lower leaves of maize (up to fourth) on which most of eggs of *C. partellus* are laid, reduced damage to some extent.
- As this practice may not be practicable in a large scale, they recommended earthling-up of the crop to bury the lower leaves as a possibility for reducing initial infestation.

Plant Resistance

- Intensive screening to identify resistance sources against (C. partellus yielded an important germplasm source. Antigua Gr. 1 and few breeding lines.
- Sources of tolerance to S. inferens have been identified. These materials are currently being used in Directorate of maize Research and also in other centres breeding programs.
- At this time, however, stem borer resistant cultivars/hybrids have yet to be released in the semi arid tropics. Further more,

the work on the screening for resistance is being done extensively and continuously and there is every possibility of identifying sources of resistance to borers and development of resistant hybrids/varieties.

Biological Control

- Many natural enemies have been reported to attack *C. partelllus*.
- Among the parasitoids reported, *Trichogramma* spp. as egg parasitoids and *Aphanteles favipes* and *T. Chilonis* and larval parasitoids are important.
- *T. cllilonis* has been found parasitising egg masses of *C. partelllus* upto 70%. *A. flavipes* is the most dominant and most widely distributed larval parasitoid of *C. partellus* in India with a parasitisation ranging from 5.3-42.8%.
- Even though a number of natural enemies have been reported on maize borers, but in general, the efficiency of natural enemies in particular farming environments is not known.
- The scope for successfully controlling maize stem borers with natural enemies is limited by the short cropping period and the lack of continuous habitat for natural enemies.

Chemical Control

- Seed treatment with carbofuran 35 ST dust.
- Seed treatment with carbofuran (40F)@ 2.5g/kg of seed was also found effective against *S. inferens*.

Effective Control Schedule

- 1st Foliar application of Endosulfan 0.1% at 15 days after sowing
- 2nd application of 4% granules al 15 kg/ha a fortnight later against *C. partellus*.

Shoot Fly: Atherigona Soccata, A.Naqui

- Very serious pests of maize in South India. In North India maize crop is not effected much except spring season in tarai area.

Nature of Damage

- The attack is maximum when the crop is in seedling stage.
- The tiny maggots creep down under the leaf sheaths till they reach the base of the seedlings.

- After this they cut the growing point or central shoot which results in the formation of characteristic dead hearts.

Control Measures

- Water logging condition should be avoided.
- Application of 10% Phorate granules in furrows before sowing @15kg/ha should be done and the insecticide should be covered with thin layer of soil after which seed should be sown.
- Spraying with 0.01% Metasystox after one week of germination.

Shoot Bug: Peregrinus Maidis

- Occasionally serious pest on maize.
- Found attacking maize and sorghum in South India and Madhya Pradesh with peak activity during August-October.
- Adult is yellowish-brown to dark brown with translucent wings in macropterous forms while brachypterous forms have underdeveloped wings, nymphs are yellowish and soft bodied.

Nature of Damage

- Adults and nymphs suck sap, resulting in unhealthy, stunted and yellow plants.
- leaves wither from top down words and plants die if attack is severe.
- Honey dew excreted by the insects causes growth of sooty mold on leaves. Mid ribs of leaves turn red due to egg laying and may dry subsequently.

Control

- Predators of coccinellid predators, C*occinella septumpunctata* and *Menochilus sexmaculata* if available will control the pest
- If Sufficient population of predators is not available spraying any systemic chemical like monocrotophos or dimethioate at 0.1% will control the disease.
- 15days after sowing spray 0.1% Endosulphan.
- After 15days of this spraying, 15kg 4% endosulphan granules per hectare are to be applied.

Integrated Pest Management And Its Difficulties

- The individual control measures may not be sufficient to adequately control maize stem borers and it is necessary to integrate the available methods.,

- In the integrated management of maize borers host plant resistance and cultural practices should be major components. The IPM recommendations are that they are too complicated to be adopted by farmers.

Difficulties Encountered in IPM

- Recommendations for stem borer management although appear promising have not carried for beyond the research stations.
- Farm sanitation can either not be improved or only be improved at high cost.
- Sowing dates are confined by several constraints and likely to be well timed in traditional farming systems that have evolved over long periods.
- Rotations are often obstructed by the vagaries of weather.
- Stem borer control through consistent intercropping is to be standardized.
- Biological control is insufficient.
- Chemical control even though efficient but limited by economical constraint.
- Integrated stem borer management finally is likely to be severely constrained by the limited management capability of farmers.
- Ever since the advent of stabilized Agriculture, storage of produce has remained an issue of utmost concern.
- Unless the problem of storage is solved satisfactorily the problem of hungry millions may continue even with substantial increase in production.
- There are a number of estimates of post harvest losses. Most conservative estimates for the post harvest losses in food grains in India even put at about 10%, a quantity good enough to feed at least 60 million people.
- Out of the total production, about 70% is retained and stored by farmers for consumption, as seed, feed and payment of wages and only about 30% is marketable surplus.
- Insects are responsible for enormous spoilage in storage they feed on grain, bore the kernel, destroy the germ portion, cause heating and deterioration in stored produce.

- As per the Pause committee report, the storage losses due to insect pests of stored grain are 2.55%. In addition they are also responsible for the qualitative losses in food grains.

Sources of Infection

- Field infestation, insects fly from stores to fields and lay eggs upon the maturing grains.
- These eggs hatch out to larvae in favourable conditions when grains reach the stores. They can crawl/fly to fresh stocks and infest them is called cross infestation.
- It has been a general practices, specially with farmers, that they keep without cleaning the empty food grain bags for use in the next season only.
- Eggs and larvae of insects remain hidden and feed on the grain fragments in the off season. Whenever the grains are filled up in such bags, infestation reappears.
- The insects, which may be present in the joints or corners of the carriers, migrate to the food grain lots and cause infestation.

The Following Methods can Help in Safe Storage of Grains

- Preventive measures
- Curative measures

Preventive Measures

- Cleaning the storage structures, sealing of cracks, crevices and holes present in the floors. Cleaning of the Sheller before their use.
- Storage structures/godowns/gunny bags should be disinfected with approved residual insecticides preferably with Malathion 50% EC, with dilution 1: 100.
- Proper stacking of the filled bags for proper hygiene and sanitation prevents insect damage in the godowns.

6

Micropropagation in Plants

Micropropagation is the practice of rapidly multiplying stock plant material to produce a large number of progeny plants, using modern plant tissue culture methods. Micropropagation is used to multiply novel plants, such as those that have been genetically modified or bred through conventional plant breeding methods. It is also used to provide a sufficient number of plantlets for planting from a stock plant which does not produce seeds, or does not respond well to vegetative reproduction.

Methods

Establishment

Micropropagation begins with the selection of plant material to be propagated. Clean stock materials that are free of viruses and fungi are important in the production of the healthiest plants. Once the plant material is chosen for culture, the collection of explant(s) begins and is dependent on the type of tissue to be used; including stem tips, anthers, petals, pollen and others plant tissues.

The explant material is then surface sterilized, usually in multiple courses of bleach and alcohol washes and finally rinsed in sterilized water. This small portion of plant tissue, sometimes only a single cell, is placed on a growth medium, typically containing sucrose as an energy source and one or more plant growth regulators (plant hormones). Usually the medium is thickened with agar to create a gel which supports the explant during growth. Some plants are easily grown on simple media but others require more complicated media for successful growth; The plant tissue grows and differentiates into new tissues depending on the medium. For example, media containing cytokinins are used to create branched shoots from plant buds and it happens in a vegetative form.

Multiplication

Multiplication is the taking of tissue samples produced during the first stage and increasing their number. Following the successful introduction and growth of plant tissue, the establishment stage is followed by multiplication. Through repeated cycles of this process, a single explant sample may be increased from one to hundreds or thousands of plants. Depending on the type of tissue grown, multiplication can involve different methods and media. If the plant material grown is callus tissue, it can be placed in a blender and cut into smaller pieces and recultured on the same type of culture medium to grow more callus tissue. If the tissue is grown as small plants called plantlets, hormones are often added that cause the plantlets to produce many small offshoots that can be removed and recultured.

Pretransplant

This stage involves treating the plantlets/shoots produced to encourage root growth and "hardening." It is performed *in vitro*, or in a sterile "test tube" environment. "Hardening" refers to the preparation of the plants for a natural growth environment. Until this stage, the plantlets have been grown in "ideal" conditions, designed to encourage rapid growth. Due to lack of necessity, the plants are likely to be highly susceptible to disease and often do not have fully functional dermal coverings and will be inefficient in their use of water and energy. In vitro conditions are high in humidity and plants grown under these condition do not form a working cuticle and stomata that keep the plant from drying out, when taken out of culture the plantlets need time to adjust to more natural environmental conditions. Hardening typically involves slowly weaning the plantlets from a high-humidity, low light, warm environment to what would be considered a normal growth environment for the species in question. This is done by moving the plants to a location high in humidity,

Transfer from Culture

In the final stage of plant micropropagation, the plantlets are removed from the plant media and transferred to soil or (more commonly) potting compost for continued growth by conventional methods. This stage is often combined with the "pretransplant" stage.

Advantages

Micropropagation has a number of advantages over traditional plant propagation techniques:

- The main advantage of micropropagation is the production of many plants that are clones of each other.
- Micropropagation can be used to produce disease-free plants.
- Micropropagation produces rooted plantlets ready for growth, saving time for the grower when seeds or cuttings are slow to establish or grow.
- It can have an extraordinarily high fecundity rate, producing thousands of propagules while conventional techniques might only produce a fraction of this a number.
- It is the only viable method of regenerating genetically modified cells or cells after protoplast fusion.
- It is useful in multiplying plants which produce seeds in uneconomical amounts, or when plants are sterile and do not produce viable seeds or when seed can't be stored (vgr. recalcitrant seeds).
- Micropropagation often produces more robust plants, leading to accelerated growth compared to similar plants produced by conventional methods - like seeds or cuttings.
- Some plants with very small seeds, including most orchids, are most reliably grown from seed in sterile culture.
- A greater number of plants can be produced per square meter and the propagules can be stored longer and in a smaller area.

Disadvantages

Micropropagation is not always the perfect means of multiplying plants, conditions that limits its use include:

- It is *very* expensive, and can have a labor cost of more than 70%
- A monoculture is produced after micropropagation, leading to a lack of overall disease resilience, as all progeny plants may be vulnerable to the same infections.
- An infected plant sample can produce infected progeny. This is uncommon if the stock plants are carefully screened and vetted to prevent culturing plants infected with virus or fungus.

- Not all plants can be successfully tissue cultured, often because the proper medium for growth is not known or the plants produce secondary metabolic chemicals that stunt or kill the explant.
- Sometimes plants or cultivars do not come true to type after being tissue cultured, this is often dependent on the type of explant material utilized during the initiation phase or the result of the age of the cell or propagule line.
- Some plants are very difficult to disinfest of fungal organisms.

The major limitation in the use of Micropropagation for many plants is the cost of production; for many plants the use of seeds, which are normally disease free and produced in good numbers, readily produce plants in good numbers at a lower cost. For this reason, many plant breeders do not utilize micropropagation because the cost is prohibitive, other breeders use it to produce stock plants that are then used for seed multiplication. Mechanisation of the process could reduce labour costs, but has proven difficult to achieve, despite active attempts to develop technological solutions.

Plant Propagation

Plant propagation is the process of creating new plants from a variety of sources: seeds, cuttings, bulbs and other plant parts. Plant propagation can also refer to the artificial or natural dispersal of plants.

Sexual Propagation (Seed)

Seeds and spores can be used for reproduction (through e.g. sowing). Seeds are typically produced from sexual reproduction within a species, because genetic recombination has occurred plants grown from seeds may have different characteristics from its parents. Some species produce seeds that require special conditions to germinate, such as cold treatment. The seeds of many Australian plants and plants from southern Africa and the American west require smoke or fire to germinate. Some plant species, including many trees do not produce seeds until they reach maturity, which may take many years. Seeds can be difficult to acquire and some plants do not produce seed at all.

Electrical Propagation

Commonly used both by hobby gardeners and in large scale industrial facilities, an electrical propagator is a low-cost, efficient and safe method of encouraging seed growth.

Asexual Propagation

Plants have a number of mechanisms for asexual or vegetative reproduction. Some of these have been taken advantage of by horticulturists and gardeners to multiply or clone plants rapidly. People also use methods that plants do not use, such as tissue culture and grafting. Plants are produced using material from a single parent and as such there is no exchange of genetic material, therefore vegetative propagation methods almost always produce plants that are identical to the parent. Vegetative reproduction uses plants parts such as roots, stems and leaves. In some plants seeds can be produced without fertilization and the seeds contain only the genetic material of the parent plant. Therefore, propagation via asexual seeds or apomixis is asexual reproduction but not vegetative propagation.

Techniques for vegetative propagation include:

- Air or ground layering
- Division
- Grafting and bud grafting, widely used in fruit tree propagation
- Micropropagation
- Stolons or runners
- Storage organs such as bulbs, corms, tubers and rhizomes
- Striking or cuttings
- Twin-scaling.

Heated Propagator

A heated propagator is a horticultural device to maintain a warm and damp environment for seeds and cuttings to grow in. This can be in the form of a clear enclosed bin sitting over a hotpad, or even a portable heater pointed at the bin. the key is to keep the moisture in the clear bin, while keeping lighting over the top of it, usually.

Seed Propagation Mat

An electric seed-propagation mat is a heated rubber mat covered by a metal cage which is used in gardening. The mats are made so that planters containing seedlings can be placed on top of the metal cage without the risk of starting a fire. In extreme cold, gardeners place a loose plastic cover over the planters/mats which creates a sort of miniature greenhouse. The constant and predictable heat allows people to garden in the winter months when the weather is generally too cold for seedlings to survive naturally. When combined with a lighting system, many plants can be grown indoors using these mats.

Tissue Culture of Woody Plants

Much has been written and more has been said about the applications of plant tissue culture to the nursery industry. The subject is still poorly understood by a majority of plant propagators. Tissue culture is often looked upon as being practical only for propagation of those plants which are difficult to propagate by conventional methods. This notion is quickly being laid to rest. An industry is emerging in various areas throughout the country which is taking advantage of technological advances in the realm of tissue culture of woody plants. This technology suggests that propagation by tissue culture is indeed applicable to the "difficult to propagate" species, and equally as important, it may offer economic advantages for some species which are considered relatively "easy to propagate".

The intent of this communication is to briefly describe the contemporary techniques of tissue culture of woody plants, and to review the advantages and limitations of tissue culture as a propagation method.

The Methodology of Tissue Culture

The variety of techniques which can be used to get plant development in vitro (that is, by techniques such as tissue culture) is considerable and completely dependent upon the species in question. Single cells of leaf tissue can regenerate whole plants, as can shoot tips, leaf pieces, root pieces, lateral buds, or stem sections. Not all of these methods are applicable to woody plants, and indeed, not all have been applied to a commercial level to any plants.

Shoot tip culture is the method in widest use for the mass propagation of woody species. An actively growing shoot tip is surface sterilized and placed on a defined culture medium under sterile conditions. The culture medium contains inorganic and organic salts (macronutrients, micronutrients and vitamins) as well as an energy source, growth regulators, and agar)to gel the medium). If the growth regulators are appropriately balanced, the shoot tip elongates, lateral buds break and begin growth, and adventitious shoots are also produced on the stem piece. This rapid proliferation of shoots results in masses of shoots being produced from a single shoot tip. Up to a hundred shoots may be produced in as little as eight to twelve weeks from a single tip. The number of shoots produced and the rapidity of shoot proliferation varies between species, and in some cases, between cultivars of a single species. Shoots are removed from the cultures at regular intervals and a portion of the mass is replaced on fresh media to continue proliferation. The small shoots which are removed are then rooted in a separate medium, either a sterile gelled medium or a peat-perlite medium (in much the same fashion as conventional woody cuttings).

Shoots produced through tissue culture are generally easy to root, even though the same cultivar may be difficult to root by cutting propagation. The resulting rooted shoot is referred to as a "plantlet" because of its miniature size, but these can generally be grown in the greenhouse at a rapid rate and with a high degree of survival. Techniques for rooting tissue cultured shoots are currently receiving a great deal of research attention as are methods for establishing these shoots into the greenhouse environment.

The Tissue Culture Laboratory

Most propagators would probably list the expense of setting up a tissue culture laboratory among the reasons for not getting involved with this technology. A lab does cost more than a mist bed! A more severe limitation is without doubt the absence of personnel skilled in the techniques of tissue culture. An individual who has had prior experience with tissue culture methodology can successfully operate a laboratory with a bare minimum of equipment. An autoclave or sterilizer, an accurate means of weighing chemicals, a small inventory of glassware, and chemical stocks or prepackaged media are all that is required to initiate cultures. The laboratory can be enlarged, more equipment added, and additional personnel hired as the need for such arises. The most critical factors are the ingenuity and experience of the laboratory manager.

As a tissue culture laboratory becomes operational, procedures other than the actual culturing process become critical. Maintenance of disease and insect free stock plants with a low titter of bacterial and fungal contamination is critical. Also important is the care and handling of the thousands of plantlets prior to their transfer to the field or containers. The ability to grow off the plantlets ultimately will limit production, rather than any limitation due to the propagation phase.

Tissue Culture Propagation of Woody Species

Major advances in the mass propagation of woody plants have been made over the past ten years. Because of the somewhat variable response of species and cultivars to the in vitro environment, much research was (and still is) needed to define the cultural conditions required by ornamental species. As an example of the variability encountered, we have observed that seedling shoot tip cultures of crabapple produce as few as one or as many as sixty-four shoots per explant after three months in tissue culture. Similar variations in the tendency for multiple shoot production have also been noted with adult clonal material of different apple cultivars.

The number of woody plant species which have been clonally propagated through tissue culture is increasing at a rapid rate, and in fact, most commercially important ornamental species have been studied. Perhaps the best publicized of the early research on woody plants involved apples, Douglas fir, and rhododendrons. An extensive commercial scale production program for apple scion cultivars and rootstocks has developed in England and in several European countries.

Using the successes of the apple, rhododendron, and maple programs as a basis, many recent advances have been made which will facilitate the mass propagation of other woody species. While it is true that cultivar differences account for lack of success with some plants, a growing body of evidence suggests that taxonomic families behave somewhat alike in tissue culture. A large number of deciduous and evergreen Rhododendron species share common cultural requirements, the most notable of which is the requirement for the same cytokinin for adventitious shoot formation. A similar relationship exists between the trees and shrubs of the Rosaceae (including crabapples, pears, plums, and hawthorns). Most rosaceous species will likely be mass propagated on media very similar to that found to be optimal for apples.

The anomalies alluded to by the crabapple seedling responses to tissue culture and the emerging similarities between the requirements of rosaceous trees and shrubs demonstrate that much research is needed to define the conditions necessary for commercial scale tissue culture production programs. As is often found to be the case with propagation systems, those systems described as optimal in the laboratory would likely need modification to be considered economical by the industry.

Impact of the Tissue Culture Industry

As is always the case for any emerging technology, the strongest supporters for the adoption of that technology are likely those involved in its development. Having acknowledged these biases, one must now consider the advantages and limitations of tissue culture propagation, and what its economic impact on the nursery industry will be. Not every nursery will Operate a tissue culture laboratory. The trend toward specialization of nursery operations as either "propagators" or "producers" is likely to continue in the future. Operation and management of a tissue culture facility requires personnel and physical facilities which are quite different from those required by a container or field production nursery. Previously in this report where the establishment of the laboratory was considered, the relatively small initial investment in physical facilities and equipment was discussed. It should be reemphasized at this point, that the training and quality of the laboratory manager is more important than the laboratory itself, and that this expertise is not currently available among contemporary nursery propagators.

Not every nursery will Need a tissue culture laboratory. Commercial operations specializing in mass propagation of nursery crops can likely produce a sufficient number of plants to satisfy the market. As with any other technology, the success of tissue culture within the nursery industry will be determined by economic factors. Can the plantlets be produced in sufficient numbers as a cost to compare with traditional propagation? Will the propagules fit into the established production sequence in the nursery industry? Are the propagules true-to-type? These answers must be obtained on a species by species, cultivar by cultivar basis.

Care and Handling of Micropropagated Plants

The propagation of plants through tissue culture, micropropagation, was viewed at one time as a research tool to study plant growth and development, growth regulators, and organogenesis. Much useful information was gathered as a result of such research. The impact which micropropagation had on the multiplication of orchids, tropical foliage plants, and selected floricultural crops is now being felt in the nursery crops industry. Rhododendrons, azaleas, mountain laurel, lilacs, and a few deciduous trees are currently being micropropagated. Tissue cultured plants are now accepted without fear of extreme variation, poor vigor, or delay in flowering as was the prior assumption.

Advantages of Micropropagated Plants

Certain morphological and physiological characteristics of tissue cultured plants make them superior to conventionally propagated materials. Tissue cultured plants (particularly Amelanchier,

Rhododendron, and Betula) tend toward a more branched habit. This tendency results in a young plant which is less "leggy" and salable earlier because of increased fullness. Tissue cultured plants produce cuttings which have increased rooting when rooted by conventional means. This phenomenon is being exploited in the "rejuvenation" of stock plants for cutting propagation. The most obvious advantage inherent to micropropagation is that more plants are produced quicker from a limited number of stock plants (one in some cases). New introductions are available in commercial quantities in a fraction of the time previously required.

The Process of Micropropagation

A brief review of the procedure of tissue culture will clarify the need for specific care and handling. A small piece of the plant to be cloned (the explant) is removed from a healthy, well-maintained stock plant and sterilized in a dilute bleach solution. The source of tissue will vary by species, but shoot tip, leaf, stem, lateral bud, and flower tissues have been used successfully for various plants. Rhododendron, for example, will produce plantlets from flower petal tissue. The sterilized explant is rinsed with sterile water, and placed in aseptically prepared containers on a specially formulated medium.

The explant may produce shoot proliferating cultures directly by enhanced lateral bud break, or the tissue may undergo a certain period of unorganized growth (callus) prior to shoot differentiation. The pattern of growth of the cultures is principally determined by the plant growth regulator content of the tissue culture medium (the auxin and cytokinin concentration). Most cultures are established within 4 to 12 weeks depending on the species and in some instances, depending on the cultivar. A shoot proliferating culture is one which can be subdivided (subcultured) to produce divisions which will continue rapid multiplication. Estimates of the rate of multiplication vary and are affected by many factors, but the production of thousands and in some cases millions of plants a year from a single explant has been demonstrated!

Morphology of A Microcutting

The small shoots produced in tissue culture are referred to as "microcuttings". Microcuttings vary a great deal in size, length and degree of leaf expansion depending upon species, cultivar, and cultural conditions. The tissue culture environment results in the production of shoots at a continuum of stages of development, from ones just recently expanded to those with elongated stems and well expanded leaves. Recognition of the type of microcutting which roots and grows out well is one of the most critical components of a successful system. The determination of the correct sequence of media and/or transfers to optimize the production of "normal" microcuttings that root and acclimate in high percentage is important. Thousands of small, spindly microcuttings which root and grow out poorly are worthless in the commercial propagation arena.

Choice of a Rooting Method

Microcuttings are rooted by modifications of two basic methods. In vitro rooting is accomplished by transferring the carefully dissected microcuttings to a medium which is free of growth regulators or has only auxin added as the growth regulator. All manipulations are conducted under sterile conditions, the process requires excessive labor input, and media must be prepared specifically for the rooting stage. Since the cuttings are provided with minerals, vitamins, sugar as a carbon source, and root promoting auxins, many researchers feel this method will allow for rooting of even the most difficult to root plants. This method does allow more flexibility in choosing cuttings since cuttings need not conduct photosynthesis to provide an energy source for rooting. Leaf expansion is therefore less critical.

In most commercial operations, the preferred rooting method is referred to as nonsterile rooting. Microcuttings are harvested (usually under nonsterile conditions by sacrificing whole cultures, but if stocks are low, sterility may be maintained at this stage) and stuck in a clean (but not sterile) mixture of peat moss and perlite or vermiculite, or in preformed foam or peat moss plugs. Cuttings may be stuck in small plastic covered containers and rooted in the lab, or placed directly into flats which are then placed in a mist, fog, or poly tent system. The nonsterile method has given outstanding results with most species tested. Rooting is accomplished in 3 to 5 weeks, and the microcuttings can be transplanted at this stage and then acclimated or acclimated directly in the rooting flats.

Acclimation of Rooted Microcuttings

Microcuttings are very sensitive to changes in the physical environment, and the success or failure of the whole operation may hinge on the acclimation process. Microcuttings are poorly adapted for growth in the greenhouse environment at the time they are removed from the rooting containers, whether these were held in the laboratory or in the mist or fog area. Microcuttings rooted and grown in very high humidity have stomates which have been shown to be "sluggish", responding to decreased relative humidity too slowly to prevent desiccation of the cutting. This problem is compounded by the apparent lack of a well developed, waxy cuticle, the barrier to water loss over the entire plant surface. Research has shown that the cuticle is present, but not functional to the extent that exists on seedling or even cutting propagated material.

Microcuttings must be acclimated to increased light intensity in much the same manner as acclimation to decreased relative humidity. If too much leaf surface is injured by direct transferal to higher light intensity (sun-burning), the vigor of the cutting will be reduced markedly because the plant is too small to have enough stored starch to force a new flush of growth. The following protocol has been successful for a number of woody species. Rooted microcuttings are transplanted to plug trays, sectioned flats, or peat moss pots into a commercially prepared medium (usually a peat moss and perlite or peat-vermiculite medium; it is important that the medium be fine textured, similar to a seed starting mix). This operation must be accomplished quickly, and care must be taken to fog or mist the microcuttings to "water them in". Transplanted flats are carried to the mist bed just as soon as they are filled. Microcuttings which are allowed to wilt during the transplanting process show poor recovery, and poor transplant survival! The mist cycle is set for a duration of 4 or 6 seconds with the interval set at 4 or 6 minutes. Bottom heat is beneficial. The mist area must be shaded with heavy shade, with a light exclusion of at least 60%. Microcuttings are left in the mist + shade for 5 to 7 days (some species may require 2 or 3 more days if there is any sign of wilting upon removal) and they are transferred to a shade house (shade greater than 50% light exclusion) for an additional 7 days.

Microcuttings usually begin a flush of growth near the end of the time in the shade house. Modifications of this procedure can be made for fog or poly tent rooted microcuttings, but the transition from high to low relative humidity and low to high light must be gradual. Microcuttings which are forced into a growth cessation by improper acclimation procedures have low survival percentages, and tend to grow out slowly and nonuniformly.

Seasonal Fluctuations in Acclimation Efficiency

Micropropagation is unique among propagation methods in that seasonality has little influence on the production of microcuttings in the lab. Most woody species, however, do display seasonal differences in acclimation efficiency and subsequent growth in the greenhouse. Microcuttings acclimated between February and September establish well in the greenhouse and generally grow out rapidly and uniformly. Microcuttings acclimated between October and January acclimate adequately, but usually do not put on the first growth flush uniformly, and generally go into a semidormant condition which requires a cold treatment to break.

Several techniques have been used in attempting to break this cycle. HPS lighting to supplement natural daylengths to 18 hours of light per day, constant feed fertilization, and incandescent light breaks have all failed to coax lagging microcuttings into active growth. We have concluded that light intensity, rather than absolute daylength, must be governing the early growth response of the acclimated microcuttings. The current recommendation is to schedule microcuttings to grow out in the greenhouse between February and September to avoid nonuniform growth following acclimation.

Should Production Nurseries Purchase Shoot Proliferating Cultures?

The concern over whether production nurseries should purchase proliferating cultures directly from tissue culture labs has arisen on many occasions recently. Shoot proliferating cultures are a convenient, compact means of transporting cuttings, they are "clean", and are available to a much lower unit cost since the cost of the rooting and acclimation stages are not included. The current recommendation is that production nurseries unfamiliar with rooting of tissue cultured materials not attempt rooting and acclimation of microcuttings without first conducting limited trials to gain experience in the procedure.

Consultation with individuals who have developed rooting and acclimation techniques has revealed that the first few experiments usually result in less than acceptable results. Individuals not familiar with the sensitivity of microcuttings to shifts in relative humidity, light intensity, and the importance of cutting selection generally will not experience the level of success necessary to make the process commercially viable. Once the tissue cultured plants have formed the first growth flush (rooted liner), they could be handled as any transplant and would not require any increased skill level from the grower.

Micropropagation is rapidly becoming a standard method in the nursery crops industry, much as it has in the foliage plants industry. The success or failure of the technique at the commercial level is often not in the lab, but rather in the greenhouse at the level of rooting and acclimation of microcuttings. Microcuttings are very sensitive to changes in environmental conditions in the early stages, but respond just as any other propagule after the first growth flush. Production nurseries should purchase tissue cultured plants at the rooted liner stage, unless a suitable system has been developed to root and acclimate microcuttings from the shoot proliferating culture stage.

Field Transplant Survival of *Amelanchier* Liners Produced by Tissue Culture

Propagation of nursery crops through tissue culture is gaining increasing acceptance from the nursery industry. Micropropagation systems from initial explanting to acclimation of microcuttings have been developed for many species. Physiological and morphological changes which occur in micropropagated plants during acclimation from aseptic conditions to the greenhouse environment have focused on two areas.

The structure and amount of epicuticular was on the leaves increases upon exposure of microcuttings to lowered relative humidity, a change deemed necessary to lower rates of water loss from the delicate microcuttings. Concomitantly, the stomata increase in rate of response to water stress after 5 days of acclimation. This change is believed to be significant to the survival of tissue cultured woody plants. Relatively little research has been conducted to determine factors necessary to acclimation of tissue cultured liners to the field environment. Smith recommends a minimum size of 10-15 cm top growth and an 8-10 cm root ball for field transplanting of tissue cultured liners. Dunstan reported that tissue cultured apple rootstocks transplanted to the field directly from the greenhouse in early spring were susceptible to frost damage and were defoliated but regrew after about 2 months of quiescence. Dunstan's transplanting procedure for summer transplanting now involves preconditioning plants under 50% shade to achieve a "partially exposed environment into which plants are placed before full exposure". Losses of tissue cultured liners were no greater than those for seedlings using this procedure.

Relatively little research has been conducted to determine factors necessary to acclimation of tissue cultured liners to the field environment. Smith recommends a minimum size of 10-15 cm top growth and an 9-10 cm root ball for field transplanting of tissue cultured liners. Dunstan reported that tissue cultured apple rootstocks transplanted to the field directly from the greenhouse in early spring were susceptible to frost damage and were defoliated but regrew after about 2 months of quiescence. Dunstan's transplanting procedure for summer transplanting now involves preconditioning plants under 50% shade to achieve a "partially exposed environment into which plants are placed before full exposure". Losses of tissue cultured liners were no greater than those for seedlings using this procedure.

This research was undertaken to determine the effect of certain factors (container type, planting date, days of outdoor acclimation) on the transplant survival of tissue cultured plants of *Amelanchier laevis.*

Materials and Methods

Microcuttings of *Amelanchier laevis* were produced in vitro according to the method described for Hally Jolivette cherry. The growth regulator concentration in the shoot proliferation stage was 0.1 mg/1 naphthaleneacetic acid and 2.5 mg/1 benzladenine. These microcuttings were rooted in plastic-covered foil containers containing a moistened peatmoss-perlite medium (1:1 by volume). Rooting was done in the tissue culture laboratory where the temperature was 24-27°C and lighting was provided for 16 hr per day at 40μ Einsteins/M2/sec by cool white fluorescent lamps. Transplanted microcuttings were immediately placed on a shaded intermittent mist bench for 3 days (6 sec most every 6 min; light intensity, 360μ Einsteins/M2/sec). After an additional 7 to 10 days of acclimation on a shaded greenhouse bench (light intensity, 270μ Einsteins/m2/sec), plants were grown under standard greenhouse cultural conditions including fertilization at every watering with 200 ppm of 20-20-20 soluble fertilizer. Plants which received an outdoor acclimation treatment were held for various time intervals under 60% shade (light intensity, 810μ Einsteins/m2/sec) prior to planting.

Field transplanting studies were conducted on the Ohio State University Dept. of Horticulture's Lane Avenue Farm. The plot used was a Brookston silty clay loam with a previous record of heavy fertilization. The plot was irrigated as needed.

In experiment 1, plantlets approximately 15 cm tall were acclimated for 0, 5, 10, or 15 days under 60% shade outdoors. Two types of containers were used, 5.7 x 5.7 cm peatmoss pots or 4 x 20 cm Cone-Tainers (Ray Leach Cone-Tainer Nursery, Canby, OR 97013). A 1:1 (by volume) peatmoss-perlite medium was used. Field planting was done June 28, 1981. A randomized complete block design with six six-plant replications was used. Experiment 2 was a replication of experiment 1 except only peatmoss pots were used. Field planting was done August 6, 1981. A randomized complete block design was used with six six-plant replications. Experiment 3 was conducted to determine how late in the year plants could be transplanted without suffering significant winter damage. At approximately 3-day intervals beginning Sept. 6 and ending Oct. 4, 1981, ten 15 cm tall plants growing in peatmoss pots were transplanted directly from the greenhouse to the field.

In all experiments, plant height, number of leaves, and number of basal lateral branches were recorded. Plants in experiments 1 and 2 were measured monthly until Oct. 4, 1981, and again on June 24 1982, after the first growth flush ceased elongation. Winter survival was assessed in May and June 1982. All data were subjected to analysis of variance. Mean separation by Duncan's multiple range test was conducted only for effects significant at the 0.05 level.

Results and Discussion

Tissue cultured plants of *Amelanchier laevis* survived transplanting in excellent condition. When results are compared by container type across all other treatments, transplant losses were negligible in either container. Shoot growth was slightly higher when plants were grown in peat pots, but those transplanted from Cone Tainers had a higher number of basal lateral branches at the end of the first growing season.

Plants exposed to a period of outdoor acclimation under shade prior to field planting appeared to have less transplant shock. Plants acclimated for 5 to 10 days exhibited the greatest incremental increase in height after 1 month. When the experiment was repeated with an August rather than a June planting date (experiment 2) and using only plants grown in peat pots, a slight benefit to prior acclimation was observed. Increasing time outdoors under shade increased both growth in height and the number of basal lateral shoots observed after 3 months. The experiment was terminated before the long term effects of acclimation on August transplanting were determined. Survival was 100%. A final study was conducted to determine how late in the season transplanting could be done without affecting overwinter survival (experiment 3). These plants were transplanted to the field directly from the greenhouse as late as Oct. 4, 1982. When survival data were taken on May 12, 1983, it was determined that all transplants had survived the winter and begun active shoot growth.

The number of leaves per plant at transplanting had no effect on transplant or winter survival or on height growth in any of these experiments. It was common for the lower third of the leaves to abscise within 2 weeks of transplanting.

These studies have demonstrated excellent field transplant survival of tissue cultured Amelanchier laevis. Survival levels of nearly 100% can be achieved: 1) without outdoor acclimation under shade, 2) when plants are grown either in peatmoss pots or Cone-Tainers, and 3) at planting dates varying from late June to early October. A period of outdoor acclimation appears to lessen transplant shock of either June or August transplanting, but the long term (12 month) benefit to this procedure with regard to shoot growth or basal branching appears negligible.

These results should not be generalized to plants which are succulent at the time of transplanting, however. Even though the *Amelanchier* liners used in this study were greenhouse grown, they did have sturdy, woody stems at transplanting.

Micropropagation: Commercial Laboratory Production

Plant Micropropagation using African Violet Leaves

Through the use of biotechnology, desirable genetic traits can be transferred from one organism to another by transfer of DNA. In plants, the DNA of interest is transferred into the new plant by using *Agrobacterium tumefaciens*, a bacterium that can infect plant tissues and incorporate part of its DNA into the DNA of the host plant. Alternatively, a particle gun can be used to directly "shoot" the DNA into plant cells. In both methods, the target for the foreign DNA is a small piece of plant tissue or a small mass of plant cells. Once the DNA has been transferred, new plants must be regenerated from the small pieces of transformed plant tissue using micropropagation (tissue culture) techniques.

Micropropagation is the aseptic culture of cells, pieces of tissue, or organs. It is possible to regenerate new plants from small pieces of plant tissue because each cell of a given plant has the same genetic make-up and is totipotent, that is, capable of developing along a "programmed" pathway leading to the formation of an entire plant that is identical to the plant from which it was derived. In addition to its biotechnological applications, micropropagation is used commercially to asexually propagate plants. Using micropropagation, millions of new plants can be derived from a single plant. This rapid multiplication allows breeders and growers to introduce new cultivars much earlier than they could by using conventional propagation techniques, such as cuttings. Micropropagation also can be used to establish and maintain virus-free plant stock. This is done by culturing the plant's apical meristem, which typically is not virus-infected, even though the remainder of the plant may be. Once new plants are developed from the apical meristem, they can be maintained and sold as virus-free plants. Micropropagation differs from all other conventional propagation methods in that aseptic conditions are essential to achieve success. The process of micropropagation can be divided into four stages:

1. Initiation stage. A piece of plant tissue (called an explant) is (a) cut from the plant, (b) disinfested (removal of surface contaminants), and (c) placed on a medium. A medium typically contains mineral salts, sucrose, and a solidifying agent such as agar. The objective of this stage is to achieve an aseptic culture. An aseptic culture is one without contaminating bacteria or fungi.
2. Multiplication stage. A growing explant can be induced to produce vegetative shoots by including a cytokinin in the medium. A cytokinin is a plant growth regulator that promotes shoot formation from growing plant cells.
3. Rooting or preplant stage. Growing shoots can be induced to produce adventitious roots by including an auxin in the medium. Auxins are plant growth regulators that promote root formation. For easily rooted plants, an auxin is usually not necessary and many commercial labs will skip this step.
4. Acclimatization. A growing, rooted shoot can be removed from tissue culture and placed in soil. When this is done, the humidity must be gradually reduced over time because tissue-cultured plants are extremely susceptible to wilting.

Teacher Preparation and Instruction Guide

Preparation for the plant micropropagation exercise should begin at least 24 hours in advance of the laboratory. The following supplies can be provided to the class in groups of two students:

- 2 fresh African Violet leaves
- 4 culture medium plates
- aseptic work surface/hood

- 1 aluminum foil packet containing 1 sterile forceps
- 1 aluminum foil packet containing 2 sterile razor blades (or 1 sterile #3 scalpel handle with sterile #10 blade attached)
- 2 sterile petri dishes
- 1 litre sterile water
- 1 500-ml or 1-liter jar with screw-cap lid (for disinfesting the leaf)
- alcohol or Bunsen burner
- spray bottle containing 70% ethanol (a spray bottle of the type used to mist plants or spray window cleaning solution works very well).

The following supplies can be shared by four students:

- *500-ml bottle (for sterilizing medium)
- *500-ml or 1-liter plastic or glass beaker (for preparing medium).

The teacher should have available for the entire class:

- commercially-purchased water for medium preparation
- 70% ethanol (dilute 700 ml of 95% ethanol with 250 ml distilled water)
- 0.1% detergent (add 1 ml detergent to 1 litre of water)
- 10% bleach solution plus a few drops of dishwashing detergent
- pH meter or pH paper
- a few mls of the following solutions for adjusting pH:
  - o 1.0 M NaOH
  - o 1.0 M HCl
  - o 0.1 M NaOH
  - o 0.1 M HCl
- parafilm for wrapping plates
- scissors
- Sharpie marking pens
- 5-10 extra sterile scalpel blades
- 5-10 extra sterile petri plates
- dishwashing detergent

Preparation of Plant Tissue Culture Medium

The culture medium is formulated so that it provides all of the compounds needed for plant growth and development, including certain compounds that can be made by an intact plant, but not by an isolated piece of plant tissue. Typically, the medium contains mineral nutrients, organic compounds such as sucrose and vitamins, and plant growth regulators (plant hormones). Agar is added if a solid medium is desired. Prepare 250 ml per group of four students. In this instruction guide, a prepackaged African violet medium from Carolina Biological Supply Company that includes all necessary components except agar and sucrose, will be used. The instructions that follow are for making 250 ml of medium. Modifications of the procedure needed to make 1 litre are included in brackets ([ ]) in bold.

Step 1. Add about 100 ml [500 ml] of deionized or distilled water to a 500-ml or 1-liter [2-liter] beaker.

Step 2. Add 1.16 grams [one package] of the prepackaged mix to the beaker.

Step 3. Add 7.5 grams [30 grams] of sucrose to the beaker. (Table sugar purchased from a supermarket can be used if reagent-grade sucrose is not available.)

Step 4. Bring the solution up to about 200 ml 00 ml] by adding deionized or distilled water.

Step 5. Stir the solution until all components are dissolved.

Step 6. Adjust the pH to about 5.6 to 5.8 with a pH meter, using NaOH and HCl as necessary. If a pH meter is not available, use pH indicating paper that can detect pH 6.0 ± 0.5 pH units. It is preferable to have the pH at slightly less than 6.0 rather than at slightly greater than 6.0. At pH < 5.0, the ability of the explant to take up nutrients will be adversely affected. At pH > 7.0, iron will precipitate out of the medium, becoming unavailable for uptake. (When using the prepackaged mix and commercial distilled water, it will take approximately 3 drops [13 drops] of 1.0 NaOH and 6 drops [4 drops] of 0.1 NaOH to raise the pH of 250 ml [1 litre] of medium to 5.6 to 5.8.)

Step 7. Bring the solution up to 250 ml [1 litre] by adding deionized or distilled water, and pour into a 500-ml bottle [1.5 to 2.0-liter bottle] to autoclave.

Step 8. Add 2.0 grams [8.0 grams] agar. (Note: the agar will not dissolve until the medium is heated.)

Step 9. Keep the Cap Loose on the Vessel Containing the Medium while Sterilizing. Sterilize the bottle of medium by placing it in one of the following:

a) boiling water for 30 minutes;

b) a pressure cooker at 15 pounds for 15 minutes; or

c) an autoclave for 15 minutes at 121*f* C (250*f* F) at 1.1 kg/cm^2 (15 psi).

Step 10. After removal from the water bath, pressure cooker, or autoclave, gently swirl the medium to mix the agar. When the agar is completely dissolved and mixed, the medium should appear clear and not turbid.

Step 11. If pouring plates, cool the medium to 45-55*f* C (110*f* F), then pour plates on a clean work surface (spray down with 70% alcohol), working near a small Bunsen burner or alcohol lamp. Use about 25 ml per plate (about one-third to one-half full). If plates are not being used immediately, put them back into the plastic sleeve, tape the sleeve shut, and store plates at 4*f* C (39*f* F).

Sterilization of Supplies

Sterilization of packets of packets of forceps and razor blades can be accomplished by wrapping each item in aluminum foil, labelling the contents with a marking pen, and:

a) baking them in an oven at 350*f* F for 15 minutes;

b) putting them in a pressure cooker at 15 pounds for 15 minutes; or

c) placing them in an autoclave for 15 minutes.

The pressure cooker and autoclave should be at the desired pressure for the 15-minute period. After the packets have cooled, they should be stored unopened at room temperature.

The students should be instructed when opening the packets to touch that part that will not come in contact with the plant material or petri dishes.

Sterilization of the sterile water needed for the rinses can be accomplished as described for sterilization of the medium.

Note: Teachers may decide for themselves which parts of the procedure can be successfully completed by the student. Middle school students enjoy transferring disinfested explants previously prepared by the instructor to the plates with medium.

Older students may be included in parts of the medium preparation and/or the disinfestation procedure. All students will enjoy seeing the growth of shoots from the leaf surface of the African violet. Seeing this regeneration process will make it easier for the teacher to start a discussion of the totipotency of plant cells and the practical application of plant propagation.

Plant Micropropagation

Aseptic Transfer of Cultures to Fresh Medium

Micropropagation is the use of tissue culture methods to propagate plants. Micropropagation if a form of clonal propagation that differs from all other conventional propagation methods in one important aspect: aseptic conditions are necessary to achieve success with micropropagation. In commercial and research laboratories, aseptic transfer is usually done in a hood that excludes particles larger than 3 microns from the work area. Contaminants of plant tissue culture are bacteria that reside on dust particles and fungi, including spores. A hood will effectively filter out all of these contaminants.

It is possible to transfer aseptically without a hood when you take all possible precautions to keep dust and spores from falling in the sterile medium. One simple, but fairly effective method for accomplishing this, is to spray a smooth counter top with alcohol and spray all other possible sources of contamination (e.g., hands, instruments, and culture vessels) before transferring. A second method used by plant pathologists is to work around a small Bunsen burner, which creates an updraft to prevent spores and dust from falling into the open dish. Extreme caution is required with this method because of the danger of burns. A third method involves spraying hands, instruments, and vessels with alcohol and working in a large new clear plastic bag. Because newly manufactured plastic is fairly sterile, working inside this sterile bag, which contains still air, will avoid contamination. An old aquarium or cardboard box lined with aluminum foil can be sprayed with alcohol and used to cut down on drafts. With all methods, it is best to keep traffic around the transfer area to a minimum.

Step 1. Put on a pair of sterile gloves. Clean your gloves with 70% ethanol initially, and reclean them if you touch any nonsterile surface or material.

Step 2. Spray your work surface with 70% ethanol.

Step 3. Spray vessels and all tool packets (forceps and razor blades) with 70% ethanol before placing them on the work surface.

Step 4. Arrange the objects on the work surface so that there is a cleared work area for the transferring process. Avoid placing any object “up wind” of your work area.

Step 5. For African violets, use a fresh, healthy leaf. The disinfestation process (removal of surface contaminants) is as follows:

a) Quickly rinse the leaf under cool tap water, then put the whole leaf into a closed container. A jar with a screw-cap lid is best. Depending on the size of the leaves, either a 500- or 1000-ml jar can be used. Wash the leaf in water with 0.1% detergent for 3-4 minutes. Gently agitate the leaf every 20-30 seconds during this washing step.

b) Rinse off the leaf and rinse out the jar with cool tap water.

c) Gently agitate the leaf in 10% bleach solution for 10 minutes. The jar should be filled three-fourths full with the bleach solution. *(The bleach solution disinfests the plant tissue material, killing most fungal and bacterial organisms.)*

d) Pour off the bleach while keeping the lid loosely in place over the container. (Be careful not to spill bleach solution on clothing because it will leave white spots after laundering.) Note: At this point, the leaf is considered sterile. All subsequent rinses should be done with sterile water, and all manipulations of the leaf performed with sterile instruments and supplies. Open one container at a time and never leave the lid off of any container longer than necessary.

e) Take the container to the sterile counter space. Spray the container with 70% ethanol before placing it in the sterile environment. Remove the lid and pour sterile water over the leaves. Fill the jar approximately half-way with the sterile water.

f) Replace the lid and gently agitate for 2-3 minutes to rinse the bleach off the leaves.

g) Pour off the rinse water in the sink, as in Step 5, d.

h) Rinse with sterile water a total of 4 times.

Step 6. Using the sterile razor blade and forceps, cut off bleach-damaged portions in an empty, sterile petri dish. With the empty dish open, hold the leaf with forceps and cut into strips 1.5-cm x 1.5-cm wide. Place 2 or 3 explants into a petri dish that contains the fresh medium. Gently press on the explants with a forceps to ensure that they make contact with the medium. Do two dishes per student.

Step 7. Replace the cover of the petri dish and wrap with parafilm. Use one 1-inch x 4-inch piece per plate. The parafilm will stretch to seal the dish. (Plant tissue cultures require much longer culture times than microbial cultures, so this additional precaution is usually worth the small extra cost and time.)

Step 8. Place the vessel under lights for 16 hours per day if possible. Keep the temperature between 24 and 26*f* C (about 75*f* F). Do not put cultures in direct sunlight, but make sure the area where they are placed is well-lighted by fluorescent bulbs. Growth is slower at lower temperatures.

Step 9. Look for small green shoots forming on or near cut surfaces. It may take 2-3 weeks or more before any visible evidence of new growth is noticeable. You may need a microscope to see these structures as they begin to grow, but after 5-6 weeks, they should be visible to the unaided eye.

If a sterile environment was not maintained, contamination will be obvious within 3 to 4 days. Materials contaminated by fungus will have a fuzzy growth on them, whereas material contaminated by bacteria will have slimy growth on them. Discard contaminated petri dishes promptly to avoid spreading plant diseases to other uncontaminated cultures.

The medium used in this exercise contains a cytokinin, and is specifically formulated to favour the production and multiplication of shoots. Once a sufficient number of shoots has been generated, portions of the explant that contain one or more shoots could be transferred to a medium that contains a higher concentration of an auxin, resulting in root production. Once roots have formed, the plantlets are transferred to pots containing a soil-based or soilless medium, and gradually exposed to conditions of lower humidity and greater light.

Rapid in Vitro Micro propagation of Sugarcane (Saccharum Officinarum L. CV-Nayana) through Callus Culture

Plant tissue culture techniques have become a powerful tool for studying and solving basic and applied problems in plant biotechnology. During the last thirty years, micro propagation and other *in vitro* techniques have become more widely used in commercial horticulture and agriculture for the mass propagation of crop plants. Sugarcane (*Saccharum officinarum* L.) is an important agricultural cash crop in tropical and subtropical region of the world and is the major source of sugar with respect to export product in many developing countries that accounts for more than 60% of the world's sugar production. It is the only member of the family *Gramineae* belong to genus *Saccharum* in which *in vitro* propagation are standardized and commercially viable. Varieties of sugarcane are highly heterogeneous and generally multiplied vegetatively by stem cutting. In tropical countries nodal sections of sugarcane with 2 or 3 nodes are commonly used as a planting material.

Lack of suitable multiplication procedure and contamination by systemic diseases are the serious problem to multiply an elite genotype of sugarcane in the open field. *In vitro* multiplication of sugarcane has received considerable research attention because of its economic importance as a cash crop. Micro propagation is currently the only realistic means of achieving rapid, large-scale production of disease-free quality planting material as seed canes of newly developed varieties in order to speed up the breeding and commercialization process in Sugarcane. As a result of which plant regeneration through tissue culture technique would be a viable alternative for improving the quality and productivity in sugarcane.

There are reports on tissue culture of sugarcane from different countries but the first attempts to regenerate plants through *in vitro* technique were made on sugarcane by Naz (2003) and Heniz and Mee (1969). Standardization of protocols for *in vitro* multiplication of sugarcane through callus culture, axillary bud and shoot tip culture have been reported by many authors. How ever, reports are scare on young meristem callus culture in sugarcane cultivar, Nayana of Orissa. The present communication demonstrates an effective high frequency regeneration method which allows for expedient multiplication of micro plants that are easily established ex vitro through callus culture of young meristem as an explant.

Materials and Methods

Explant source: Healthy young meristems were collected by removing the leaf sheath from field grown plants of sugarcane (*Saccharum officinarum* L. cv- Nayana) maintained in the P.G. Department of Botany Utkal University and brought to the laboratory. These young meristems were cut into thin smaller pieces of 1.0 to1.5 cm length.

The explants were washed thoroughly under running tap water for 20-30 minutes followed by bavistin 0.2% for 10 minutes and then washed with sterile distilled water and transferred to laminar air flow cabinet. The young meristem explants were treated with 70 % alcohol for 30 second to one minute, followed by another treatment in 0.1%

(w/v) mercuric chloride (HgCl2) for another 5 minutes. Finally, the young meristem cuttings were washed thoroughly 3 to 5 times with sterile distilled water before the inoculation in to sterilized nutrient agar media pre-packed in culture tubes. All the above operations were performed under aseptic conditions in laminar airflow cabinet.

Culture medium and condition: The young meristem cutting explants were inoculated on to sterilized semisolid basal MS medium supplemented with different concentrations and combinations of different plant growth regulators.

Callus induction: For callus induction different range of concentration of auxins are tried individually such as 2,4-D (0.5,1.0,2.0,2.5,3.0 and 4.0 mg/l), IBA (0.5,1.0,2.0,2.5,3.0 and 4.0 mg/l) and NAA (0.5,1.0,2.0,2.5,3.0 and 4.0 mg/l) with MS basal medium.

Shoot regeneration Medium: White friable calli were cultured on MS medium supplemented with different combinations and concentrations of BAP, Kinetin (0.5- 4.0mg/l) and IBA, NAA (0.1-1.0mg/l) for multiple shoot regeneration either individually or in combination.

Rooting medium: Elongated micro shoots measuring about 5-6 cm in length were excised from culture tube and transferred to half-strength (1/2 MS) MS medium supplemented with different concentrations of IBA,NAA and IAA (0.5-3.0mg/l) either individually or in combinations.

Environmental condition: The pH of the medium was adjusted to 5.8 before gelling with Agar (6g/l bacteriological grade, Qualigens , India) and prior to autoclaving for 20 min at 120 o C and at 15 lbs psi pressure.

Sugar was added at the concentration of 30 gm/l. Molten medium of 20 ml was dispensed into the culture tube and plugged with nonabsorbent cotton wrapped in one layer of cheesecloth. All the cultures were incubated in a growth room with a 16h photoperiod except callus culture (cool, white fluorescent light -30μmol m-2S-1) and the temperature was maintained at 25 ± 30C with 70-80% relative humidity in the culture room. Each treatment consisted of 10 replicates and repeated thrice. For better callus induction just after autoclaving the culture tube containing semisolid sterilized media should positioned , such a manner as if the inside semisolid media in the culture tube should spread to maximum to wards one side only. At the time of inoculation 2 to 3 explant per culture tube was used. For callus induction experiment culture tubes were kept under 20, 25 and 30°C in complete dark or light/dark condition with 16/8 h light/dark photoperiod with 140μmol m-2S-1 light from cool, white fluorescent lamps.

Acclimatization and transfer of plantlets to soil: Plantlets with well-developed roots were removed from the culture medium. Washed the roots gently under running tap water and were transferred to plastic trays for hardening which contain autoclaved garden soil, farmyard manure and sand (2:1:1). The harden plantlets in the plastic trays were covered with porous polyethylene sheets for maintaining high humidity and were kept under shade in a net house for further growth and development. All were irrigated with 1/8 MS basal salt solution devoid of sucrose and inositol every 4 days for two weeks. After 30 days the plantlets were transplanted in to the soil in field condition.

Statistical Analysis: Experiments were set up in a Randomized Block Design (RBD) and each experiment usually had 10 replicates and was repeated three times. 20 to 30 explants were used per treatment in each replication. Observations were recorded on the percentage of response of callus formation, percentage of response of shoots, number of shoots per callus, shoot length, percentage of response of roots, roots per shoot and root length respectively. The treatment means were compared using Duncan's Multiple Range Test (DMRT) at a 5% probability level according to Gomez and Gomez (1976).

Results and Discussion

Callus Induction: Callus induction was observed within two weeks, after inoculation of the explants on MS medium containing different concentrations of IBA, NAA and 2,4-D (0.5,1.0,2.0,2.5,3.0 and 4.0 mg/l). Although in all concentrations of 2, 4-D the callus induction was triggered, but more profuse callus induction was observed at 2.5 mg/l of 2,4-D with full potential of callus regeneration from the explant of the cultivated varieties, Nayana. On this media composition the explants produced creamy white callus.

The percentage of callus induction was 100 %. Such type of calli has also been reported by Khan et al., (1998) and Khatri et al., (2002). Begum et al., (1995) found that 3.5 mg/l of 2, 4-D produced highest percentage of callus induction from leaf base explant in Bangladesh Nagabari variety of sugarcane. Islam et al., (1982) also reported that 0.5-5.0 mg/l of 2,4-D showed callus induction from leaf tissue on MS medium.

The concentrations NAA at 2.0 and 3.0 mg/l produced small amount (20-30%) of callus with grayish globular and hardy in nature. On second sub-culture the calli were turned into non-regenerable callus in NAA combination media with reduced callus weight where as in IBA+ MS basal media composition explant does not showed any remarkable callusing.

The cultivated variety Nayana produced poor callus in lower concentration of 2, 4-D and did not at all in 0.5-1.0 mg/l of hormonal combination with MS basal media. All these studies indicated that the sugarcane variety Nayana require higher concentration of 2, 4-D for callus induction.

Regeneration of Micro shoots: Various concentration of cytokinin (BAP and Kinetin) and auxins, (IBA, NAA) were used in different concentration and combinations for shoot regeneration. During this investigation shoot formation was highly influenced by concentrations and type of the growth regulators used in the experiment. Among different concentrations and combinations for shoot multiplication , best performance was showed on MS medium supplemented with BAP (2.0 mg/l) + IBA (0.5mg/l).

On this combination 92 % of explant produced shoots .The number of usable shoots was 12.4 ± 1.90 with average length of the shoots 6.2 ± 0.37 cm of the cultivated variety Nayana. The second best performance was found on MS medium supplemented with BAP (2.0mg/l) + IBA (1.0mg/l) in which average number of usable shoots was 10.5 ± 1.31 with mean length of the shoot 4.0 ± 0.61cm. Islam et al., (1982) also reported the positive effects of BAP+IBA combination on shoot formation in sugarcane. It was also observed that BAP+NAA combination showed effective result but not superior than BAP+IBA combination.

However, combinations of high level of cytokinin and a low level of auxin were essential for differentiation of adventitious shoot in sugar cane young meristem callus of the variety Nayana than individual concentration of cytokinin. All these studies concluded that regeneration potential of callus was specific and genotype dependent phenomenon and at the same time it parallel with the hormonal concentration and combinations. It was also observed that callus derived from different auxins showed different regeneration potential. Callus induction, proliferation and regeneration potential in sugarcane exhibited synchrony to each other.

In vitro rooting and acclimatization: Different types of auxins were used at different concentrations and combinations to regenerate adventitious roots. Among them NAA and IBA was found to be comparatively better response than IAA for profuse rooting. NAA+IBA combination showed positive result. Best rooting was observed in 1/2 strength MS medium supplemented with 2.5mg/l NAA and the highest number roots per micro shoots were 13.4 ±1.5 , which take only 8-10 days for initiation of root primordial with average root length 4.0 ± 0.94 cm for the variety Nayana found in Orissa.

According to Lal and Singh (1994) root can be easily induced on culture shoots by their transfer to another medium with or without NAA ,where optimal growth were observed with 1/2 strength of MS medium . Baksha et al., (2002) used 5.0 mg/l NAA for best rooting response in half strength MS medium. Sabaz et al., (2008) used 1.0 mg/l IBA as the best root initiating growth hormone with highest number of 41 roots per plant. Gosal et al.,(1998) obtained rooting on liquid MS medium containing NAA (5 mg/l) and 70 g/l sucrose. Ali and Afghan (2001) observed only 6 - 7 roots after 3 weeks on MS medium containing 2.0 mg/l IBA and 6% sucrose. Baksha et al., (2002) also got rooting response at 0.1 - 0.5 mg/l IBA along with 0.5 - 2.0 mg/l BAP but these were of poor quality. These findings also agree well with the previous findings of Nadar and Heinz (1977).

Alam et al., (2003) reported best rooting response at 2.5 mg/l IBA with 16 number of roots/explant having 1.1 cm root length. Mamun et al., (2004) obtained best results of rooting on MS medium supplemented with auxins (NAA + IBA) 0.5 mg/l for each one. We also found that 0.5 mg/l NAA+2.5 mg/l IBA was the second best feasible rooting response with 11.3±1.08 number of roots and 3.7 ± 0.47 cm of root length. The plantlets with well developed shoot and roots after acclimatization were successfully transplanted in soil with 85% acclimatization of survivability potential. It is difficult to release a new variety of sugarcane by the conventional breeding method for the genetic behavior of sugar cane. More over also it takes long times to release a stable variety. Thus tissue culture technique can plays an important role in this regard for supply of disease free quality planting material in a year round basis and true to true types of the mother plant.

7

Prospects and Limitations of Agricultural Biotechnologies

Introductory Remarks

It is now generally that the practice of plant breeding will move forward as progress is made in knowledge and technology. Breeders in the future will not only be forced to hone their traditional skills but will also have to integrate them with knowledge and experience to coincide with more recent advancements and technologies derived from cellular and molecular concepts and approaches.

For worldwide crops such as potato, rice, maize and in the future wheat, as well as for high value vegetable crops, the future of breeders depends upon the use of a combination of breeding practices (traditional breeding, genetic engineering and tissue culture). In addition to an increasing reliance on breeding, agriculture will also depend on biocontrol to complement and to make possible the use of chemicals that are compatible with intelligent management of the natural resources essential for a sustainable yet highly productive agriculture.

Thus most aspects of agriculturally relevant biotechnology are covered. Plant tissue culture techniques not only provide essential ways for the clonal propagation of many agriculturally important crops(e.g woody plants ornamentals and vegetables) but they are also the basis for the production of transgenic plants and are at the forefront of recent studies in plant physiology, developmental biology and biochemistry. Spectacular progress in our knowledge of plant sciences, including those relevant to agriculture and breeding, can be expected from the integration of more traditional sciences with mole-cular and cellular concepts and techniques. A prime example is seen in the impact molecular genome analysis is already having on traditional breeding (via so called map-based breeding) but also on research and molecular breeding (e.g, via map based gene-cloning) This rosy description of the future of genetic engineering is richly tempered in this book by a realistic assessment of the many problems that still need to be faced before what is predicted can be accomplished. It was therefore wise to ask scientists well known for their "no-nonsence" approach such as Ingo Potrykus and his colleagues to write about genetic engineering of crop plants.

To quote from their introduction: " it has been possible to develop the state of the art of gene transfer to plants to a level at which many of the major crop plants are accessible to the technique and for which we have good reason to believe that there is no basic biological principle that will prevent gene transfer to a specific crop plant " Yet these authors clearly identify the considerable research and development that is still needed before one can, in many but not all instances, consider large scale applications.

A recurrent point made and well documented in various chapters of this book is that genetic engineering-because it is not limited by the natural barriers that ensure the maintenance of species and prevent transfer of genes between unrelated organisms-is an extremely powerful way to increase the genetic variability available to breeders. The present rise in interest to acquire new knowledge via fundamental research can best be illustrated by reminding ourselves of recent progress in our understanding of the molecular physiology of development, of the mechanism of action of old and newly discovered phytohotmones and of growth signals used by microorganisms that are pathogens or symbionts of plants (such as for instance lipo-chitooligosaccharides). Biotechnology can contribute to making traditional agriculture more productive while at the same time more sustainable. It can also help achieve the goals of "non-food" and or cash crops agriculture as has been described in the various chapters dealing with crop improvement and metabolite production. Hopefully sufficient attention will be paid to regional crops of importance by local populations in developing countries.

Finally topics such as "Biofertilization" and "Bioremediation" are convincingly dealt with Topics such as "Biotechnology of Farm Animals" and "Marine Biotechnology" as well as the unavoidable but all important "Legal and Public Aspects" round out the all inclusive title of

Agricultural Biotechnologies. Some topics have already come to blossom and illustrate well the potential of present day plant sciences. There is the breakthrough in the cloning and characterization of resistance genes along with significantly improved understanding of the nature of avirulence genes and elicitors. Many hypothetical models predict that resistance genes should be involved in the signal transduction pathways linking pathogen derived signals with the regulated expression of defence genes and of programmed cell death (hypersensitivity reaction). The first available data on resistance genes show that this concept may well be correct in its very general sense, even if all resistance genes do not code for elicitor receptors.

We may safely predict that in the near future we will know more details of the molecular mechanisms underlying plant-pathogen resistance mechanisms. Several elicitor receptors will soon be isolated and characterized along with genes involved in localized pathogen-induced programmed cell death. Plant breeders will make use of isolated resistance genes to produce crops with improved resistance to a variety of pathogens and pests. The content of this book fully justifies the conclusion that Biotechnologies are bound to play an ever growing role in agriculture not only in the long term future, but also presently, and in the near future.

Fortunately the progress in applied plant biotechnology corresponds with and is in fact stimulating fundamental scientific progress. This is important because the real driving force behind this science is the success of applied plant biotechnology such as: 1) plant protection based on the expression of several different introduced genes coding for different principles acting on the same targets and thereby providing a possible solution to the problem of the emergence of resistant pathogens and pests, 2) new quality traits as well as future crops producing tailor-made non-food products (lipids, carbohydrates, and biodegradable thermoplastics). Only if plant biotechnology becomes a commercial as well as an environmental success will sufficient support be available for this fascinating research.

Plants and Agriculture

Human population has grown steadily since the advent of agriculture, nearly 12,000 years ago, which ensured a continuing and reliable supply of food. The most rapid growth occurred during the past two centuries during which time agriculture became highly industrialized, public health improved and there were increases in incomes and international trade in food grains. Despite the age-old human concerns about balancing population and food supplies and many population control measures the worlds population is projected to continue to grow into the early decades of the 21st century, stabilizing at 9-11 billion sometimes during 2030-2050. Much of this growth will occur in the already overpopulated underdeveloped and poorer regions of Africa, Asia and Latin America, which will be home to nearly 90% of the human population.

Food shortages were common during the early and middle years of the 20th century. The introduction of Green Revolution varieties of wheat, rice and maize during the mid-1960s reversed this trend and helped avoid major food shortages by keeping increases in food productivity slightly ahead of population growth. During this period, both China and India, the two most populous nations with chronic food shortages, became bet food exporters. Realistically however such increases in food productivity can not be sustained indefinitely. It is not surprising, therefore that increases in food productivity have begun to decline during the past few years. Furthermore, improved economic conditions in China and India have created greater demand for better and varied food products, particularly poultry and meat, which require greater supplies of feed grains. As a result, both China and India have now become net importers of food. World food reserves have declined from a high of 77 days to less that 50 days.

With the current trends of population growth and agricultural production the demand for food in the most populous parts of the World will double by the year 2025, and nearly triple by 2050, Increases in food productivity of this magnitude can not be brought about in such a short period of time by conventional breeding, especially when some of out most important crops are approaching the physiological limits of productivity, not by increasing the amount of arable land, which accounts for 97% of all food production in the world..

Arable land, which is finite and comprises about 3% of the earth's surface, is deteriorating and decreasing as a result of soil erosion, salinization, over cultivation, and acidification. These factors, combined with expected increases in population, will actually decrease the global per capita arable land from the current 0.28 hectare to 0.17 hectare by the year 2025. In addition, fresh water supplies, essential for modern high-input irrigation agriculture, are becoming limited by increased human and agricultural use, and polluted by agricultural run-off and widespread use of agrochemicals. It is feared that the resulting food shortages in the overpopulated parts of the world during the 21st century may lead to widespread social, economic, and political unrest, making food security the single-most serious threat to international peace and security. At the same time it is well known that countries with efficient agricultural systems generally have high living standards, strong economies, lower rates of population growth and democratic forms of government Increasing food productivity in a sustainable manner will, therefore, not only provide adequate nutrition to the expanding humanity, but will reduce population growth, protect the environment, promote economic development and ensure social and political stability.

The challenge for the agricultural sector during the next few decades is therefore clear double food production by 2025, and triple it by 2050, on less per capita land, with less water, under increasingly challenging environmental conditions. The situation is further complicated by the fact that in spite of the heavy use of agrochemicals, modern agriculture still loses nearly 42% of crop productivity to competition with weeds and to pests and pathogens, and an additional 10-30% to post-harvest losses due to a variety of factors, especially in the developing countries where storage conditions are poor. During the 20th century traditional plant breeding has brought about enormous increases in crop productivity. However, plant improvement by hybridization is slow, and is restricted to a very small gene pool owing to natural barriers to crossability. Beginning in the early 1980s, advances in plant cell culture can genetic transformation have overcome these barriers by making it possible to transfer defined genes into all major food crops, including cereals, legumes, cassava, potato, and many vegetable and fruits. The entire global gene pool-whether it be plant, animal, bacterial or viral-is available for utilization. The first genes that have been integrated into crop species provide resistance to non-selective and environment friendly herbicides, and many pests and pathogens. Increasingly large acreages of transgenic maize, soybean, potato, tomato and cotton are being commercially grown for human use and consumption. Carefully planned introduction of such crops on a worldwide scale would greatly help in reducing or even elimination the enormous crop losses attributed to weeds, pests and pathogens. The use of such crops will also have a beneficial effect on the environment by significantly reducing the use of agrochemicals. Other genes for improving crop productivity, and manipulating starch/protein/ oil quality and quantity, resistance to environmental stresses such as temperature and drought are also being isolated and studied. In the foreseeable future, these will be used to produce second generation transgenic crops. It is clear that during the next few decades a wide variety of transgenic crops will become integrated into agricultural systems in the industrialized countries, Introduction of such crops into the developing countries, which need them most, will be slow and largely through the efforts of multinational biotechnology companies, because most of the developing countries currently lack the scientific and industrial infrastructure to develop and introduce these technologies into their agriculture.

In the long term however, much of the increase in food production to meet the dual challenge of population growth and food demand must occur in the developing countries. This will impact greatly on their overall economic development, which will help control the relentless increase in the population. It is critical, therefore that scientific and technical manpower and infras-tructure be created in developing countries to take scientific and technical manpower and infrastructure be created in developing count-ries to take advantage of the remarkable advances in agricultural biotechnology.

Assistance should be provided to the developing countries be international organizations such as the United Nations Educational Scienticfic and Cultural Organization (UNESCO), the Food and Agriculture Organization of the United Nations (FAO) and the World Bank through its many international agricultural research centres, like those in Mexico and the Philippines, where Green Revolution originated, International agriculture in the latter half of the 20th century was dominated by the Green Revolution, Considering the power and potential of agricultural biotechnology, it is likely that the Gene Revolution will dominate the agriculture of the 21st century. Like the Green Revolution, it may help accelerate the rate of food productions, save lives from hunger, create livelihoods in rural households, save large tracts of land that would otherwise be needed for food production, reduce birth rates, ensure low food prices, stimulate broad-based economic growth, expand world trade, and help create a sustainable agricultural system for future generations. Finally, it should be understood that although plant breeding has long been, and will continue in the future, to be indispensable for plant improvement, it must now be complemented and supplemented by molecular breeding and genetic transformation, in order to establish a sustainable agricultural system for the 21st century.

Animal Biotechnology

Animal Biotechnology has developed rapidly from the early 1980s when the first transgenic mice and first in vitro produced bovine embryos occurred. Today animal breeding companies are using marker assisted selection to provide earlier and improved selection of breeding animals. Gene mapping efforts are annually identifying a large number of candidate genes for testing of usefulness in marker assisted selection. The propagation of genetically valuable or transgenic animals is enhanced by in vitro production of embryos.

There are already more than a dozen commercial companies selling or producing for customers in vitro produced embryos. Cloning of embryos of cattle and sheep has been shown to be possible, but low efficiencies have thus far prevented commercial use. Animal producing and genetic selection systems can be made more efficient by sexing the sperm used for in vitro production of embryos. The flow sorting of sperm cells to separate X from Y bearing cells has been successful in most species tested and in cattle and swine has resulted in offspring of the desired sex. While effective with in vitro embryos production, the sexed sperm are of less than normal viability and sperm numbers and not yet useful in artificial insemination or the freezing of sperm.

The artificial insemination and embryo transfer industries are today heavily dependent on use of cryopreserved sperm and embryos. However the cryopreservation of oocytes and the combination of some biotechnologies such as embryo biopsy and sexing with cryopreservation are still in the research stage. Transgenic cattle, sheep swine and poultry have been produced primarily by microinjections of DNA into pronuclei of eggs or by use of viral vectors. The intended use has been to change animal growth, disease resistance, wool production, or to produce new products in milk. The efficiencies thus far have been much lower than in mice. Therefore there has been little commercial use of transgenic technology except for the involvement of several companies in production of pharmaceutical products in milk.

The use of cultural embryonic stem cells and nuclear transfer to make offspring has resulted in calves and lambs. This technology offers promise of making large numbers of nearly identical offspring or more efficient gene transfer or selection. However thus far the technology is not sufficiently efficient for commercial use. Growth hormones and growth promoting factors are being developed for improved efficiency of animal growth. Growth hormones are commercially in use at present for promotion of milk production in cattle. The anatomy and physiology of birds and the late stage of egg laying prevents the introduction of DNA into pronuclei as in manuals. Thus other methods such as oocytle microinjection of DNA, viral introduction of DNA and introduction of DNA into primordial germ cells are being used in avian gene transfer research. Oocyte microinjection, blastoderm transfection and primordial germ cell manipulation have all been successful in producing transgenic birds. The production of transgenic birds appears possible but is not yet in commercial use. Patented applications of recombinant DNA biotechnology to animal agriculture became apparent in the early to mid 1980s with development of recombinant vaccines effective against calf scours and against rabies. Today numerous vaccines in use are produced by recombinant technology. Biotechnology has also revolutionized diagnostic techniques by introducing higher sensitivity and specificity. Monoclonal antibodies, DNA probes, Southern and Western blotting, DNA and RNA amplification through PCR, constitute as integral part of the diagnostic arsenal at the disposal of the advanced laboratories.

Based on these techniques rapid animal-side field test used by the clinicians furnish results in real time, permitting immediate decisions as to the proper treatment, vaccination, or stamping-out policy. High technology based diagnosis is expected to become a routine in every diagnostic laboratory forming a sound basis for control and treatment of animal diseases.

In the short time of approximately fifteen years, animal biotechnology has moved from an unknown and unstudied science to a strong discipline of science and extensive commercial application. The science and its commercial use are rapidly progressing. The ultimate benefactor will be humans through improved food supply, nutrition and health care products available to humans around the world.

Biotechnology in Agriculture

During the last decade, tremendous progress has been made in the area of plant cell, tissue & organ culture. In vitro techniques constitute an important component of biotechnology, and have the potential not only to improve the existing cultivars, but also for the synthesis of novel plants & early release of high-yielding plants resistant to various diseases, pests, stresses & temperature. Green revolution of recent years has been the result of efforts that led to the improvement & better selection of species of agronomic interest.

Some specialists feel that biotechnology is the second phase of green revolution, studies are already in progress to obtain crop plants having higher photosynthetic efficiency, improved fixation of atmospheric nitrogen, increased nutritional quality & greater tolerance of salinity, alkalinity, acidity, deficiency of nutrients & stresses. Wide hybridization through embryo & ovule culture both at the interspecific & intergeneric levels have been achieved in a number of crops through the culture of the hybrid ovules & embryos.

In vitro fertilization techniques have also been employed to overcome incompatibilty. Haploids produced through anther culture & their diplodization have resulted in the release of varieties in crops such as wheat & rice.

Biotechnology

The term biotechnology is composed of two words-bio (Greek-bios-means life) and technology (Greek technologia-means systematic treatment). In other words, it is the science of applied biological processes. Accordingly to the oxford Dictionary, technology is the "Scientific study of the practical or industrial arts". Spinks (1980) defined biotechnology as, "the application of biological organisms, systems or processes to manufacturing & service industries". European Federation of Biotechnology (1981) defined biotechnology as, "the integrated use of biochemistry, microbiology & engineering sciences in order to achieve technological (industrial)application of the capabilities of micro-organisms, cultured tissue cells & parts thereof". In the report Biotechnology: A Development plan for Canada (1981) biotechnology is defined as "the utilization of a biological process, be it microbial, plant or animal cells, and their constituents, to provide goods & services".

Production of Pathogen-free Plants.

Plants traditionally being vegetatively propogated are systemically infected with viruses & other pathogens which greatly reduce yield & also Q uality of the marketed commodity. Virus diseases like potato leaf roll virus (PLRV) or potato virus Y (PVY) for example, can cause upto 95 % reductions in the tuber yield of potato crops.

Increase in yield upto 300% (Averaging 30%) has been reported following replacement of virus-infected stock with specific pathogen free plants. Since majority of viruses infect plants in a systemic manner their elimination may be achieved through meristem tip culture. Plant species for which virus-free plants have been obtained.

Species Virus Eliminated

1. Allium sativum (Liliaceae) Garlic mosaic virus
2. Brassica Oleracea (Brassicaceae) Cabbage black ringspot virus, Turnip mosiac virus, cauliflower mosiac virus
3. Chrysanthamum sp. (Asteraceae) Chlorotic mottle, Green Flower, stunt vein mottle, virus B.
4. Dahlia Sp. (Asteraceae) Dahlia mosaic virus, Tomato Aspermy, vein mottle, virus B.
5. Glycine max (Fabaceae) Soyabean mosaic virus. Impomoea batatas Feathery mottle, Hanmon Mosiac, (Convolvulaceae) ' Rugosa mosaic
7. Lilium Sp. (Liliaceae) Cucumber mosia virus
8. Monihot Sp. (Euphorbiaceae) African cassava mosiac
9. Musa sp. (Musaceae) Cucumber mosiac virus
10. Nicotiana tobacum (solanaceae) tobacco mosaic virus
11. solanum tuberosum(solanaceae) potato virus.

Although the apical meristems are often free of viruses, this cannot be regarded as a phenomenon of u niversal occurrence. The success of meristem tip culture depends upon several factors. One of the most important factors is the relative distribution of viruses in the growing tip of donor plants. There are some viruses which invade the growing tip viz, TMV, potato virus X, cucumber mosaic virus. In such cases virus free plants are obtained by combining meristem-tip culture with high temperature treatments.

Production of Disease Resistant Plants

This is one of the most useful applications of tissue culture in crop improvement. In potato, somaclones have been screened for both late & early blight resistance. In maize, somaclonal variation has induced resistance to race T of southern corn leaf blighrt. In sugarcane, resistance to diseases like fiji & downy mildew have been recovered. In lucerne, selection of cell lines & plants resistant to the toxin of *Fusarium oxysporium* has been accomplished.

Improvement of Nutritional Quality

One of the major sources of protein for human & animal commnsumption is constituted by the proteins contained in seeds of many plant species. The cereals & legumes which are major sources of storage seed proteins, contain limited amount of certain amino acids which are essential for human beings. Majority of these cereals are deficient in lysine whereas legumes are deficient in sulpher amino acids. A wide range of approaches have been employed for improving nutritional quality of various crop plants. Important among them are selecting cell lines resistant to amino acid or their anologues of lysine, tryptophan, proline & phenylalanine.

Isolation of variants over producing specific amino acids in culture has been successful, but expression in the whole plant & especially in the seed has not yet reached the level required to make an impact on protein quality.

Selection for Salt & Draught Tolerance

Continued efforts to increase intensity of cropping for increasing production from limited land resource by extending irrigation facilities have resulted in the gradual build-up of salt concentrations in the soil. This has resulted in loss of productivity on such soils. Salt tolerant lines have been produced in crop plants such as tobacco, tomato, cereals. Salt tolerance has been incorporated into rice lines with improved plant type. Some of the improved rice cultivars viz., IR 42, IR 43 & IR 52 are salt tolerant.

Development of cultivars tolerant draught can contribute significantly in agriculture economy. Tolerance to draught is a polygenic trait & involves highly complex osmo-regulatory functions. In tissue cultures, simulated draught conditions have been achieved through incorporation of non-penetrating osmotic solutes such as PEG (polyethylene glycol) and dextrans in the media.

Production of Genetically Variable Plants

The success of any crop improvement programme depends on the usable genetic variability in the base population cells in culture offer an excellent systems for inducing variations & regenerating pure mutant types. Genetic variation can be an option to lessen our reliance on cost intensive germ-plasm collection & conservation programmes.

Variant producing capacity of cell culture can be augmented to a great extent by employing physical & chemical mutagens. Somaclonal variation has been extensively exploited for the improvement of a sexually propogated crops viz., potato, sugarcane. From cell cultures, some superior cultivars have already been produced in sugarcane which are high yielding, drought resistant & temperature tolerant.

Biofertilization

Molecular nitrogen in the atmosphere is converted into Biologically converted forms by nitrogen fixing micro-organisms e.g. Rhizobium. The most sophasticated approach to biofertilization is to create plants that possess the genetic capacity for nitrogen fixation. Attemps are being made to transfer genes for nitrogen fixation (hifgenes) from bacteria to plants.

Rapid Clonal Propagation

Tissue culture has found its best commercial application in production of cloned plants at a very high rate as compared to conventional methods. It is important specially for initially building up of propagation stock of elite clones or individual plants which are otherwise slow to multiply. A number of agriculturally important plants have been clonally multiplied.

Genmplasm Storage

The primitive cultivars & wild relatives of crop plants constitute a pool of genetic diversity which is invaluable for further breeding programmes. There are over 20,000 plant spices which are rare or threatened with degradation of their neutral habitats. The most economical form of storing germ plasm for seed propagated species is seeds. However, there are certain limitations of this method. Therefore, various methods of invitor storage of germplasm are thus of great practical significance for long term storage of germplasm. Presently, there are two approaches to invitor germ plasm storage. Slow growth technique & cryopreservation.

Biological Control of Agricultural Pests

Insects consume nearly one-third of human food supplies on earth. One of the major goals of biotechnology is to develop target specific biological pesticides that will kill some specific pest but will not harm other species. Several insect-specific pathogens, developed through biotechnological processes, are being produced commercially for their use as microbial pestices.

Bacillus thuringienis inocula have been used as bio-insecticides. Toxin produced by the bactenium kills the gypsy moth. The prevention of crown gall disease formation is achieved by spraying young plants or seeds avirulent species *Agrobacterium radiobactor* var, *radiobactor*. Microbial insecticides have a number of advantages over traditional chemical insecticides:

a) they are specific for a small number of species of insects & do not kill plants, animals & beneficial insects.

b) they are cheaper than organic pesticides and

c) they do not leave any residue effect.

Selection for Herbicide Resistance

Herbicides are used to control weeds in agricultural fields. Tissue culture techniques offer several advantages in developing herbicide-resistance plants:

1. The techniques for selection are relatively rapid;
2. Small volumes of cells represent the genetic potential of hectares of fields.
3. Metabolic studies are facilitated by uniform & sterile conditions &3
4. Possible new types of resistance not found in nature may be obtained from plant cell cultures.

Attempts to produce herbicide tolerance using somaclonal variation & invitor selection have already been made.

Biotechnology and Indian Agriculture: The Challenge of the Next Millennium

The world has won the important battle in the area of food security, but the war is still on. A total of 800 million people, that is one out of every six persons, in the developing world do not have access to food. One-third of all pre-school children in the developing countries are food insecure. We, thus, have a big challenge ahead of us as we enter the next millennium. It is true that the mass starvation that was predicted for Asia in seventies and eighties did not occur. It is only because Science was effectively put to work to raise agricultural productivity. In India, the ' *Green Revolution*' was a success due to the introduction of improved seeds, fertilizer, irrigation and plant-protection measures combined with positive policy support, liberal public funding for agricultural research and development, and dedicated work of farmers.Notwithstanding all-round achievements, the basic problems of food security, poverty, equity and sustainability, continue to be a cause of concern in India today.

The Indian Challenge

There are several concerns about Indian agriculture that we need to address. Even in the green revolution crops, rice and wheat, the

Indian average yield ranks around 50th in the world. A weak agro-based industry, the problems of weak marketing, storage, transportation, credit support, etc, haunt us. Only 0.3% of agricultural GDP is spent on agricultural R &D. The population is growing at an alarming rate of 1.8% per annum. It will be around 1.3 billion by 2020 and would require 50% additional foodgrains. A total of 200 million people are still below the poverty line and have insufficient access to food. Declining resource base, degeneration of natural resources, including soil fertility, water environment, etc, are some other concerns.

Food production systems centre around a few states only. Out of 17 big states, only 7 are self-sufficient and surplus in foodgrains. There was a significant decline in per capita intake of calories and protein between 1975 and 1990. Inspite of all these problems, there is plenty of good news. India harbours a vast diversity of basic natural resources. Tremendous bio-diversity of agriculturally important micro-organisms exists, which are not yet been harnessed. We have a variety of crops, including quality fruits, vegetables, foodgrains, and a number of plantation and commercial crops. Indian fisheries resource has spread over 8,100 km of coastline on the east and west, and provides access to marine resources and transportation.

Very large proportion of area of various food crops is under low productivity category. It can be increased with inputs of high class science. There is a scope to cultivate sizeable portion of waste lands through soil amendment and introduction of agroforestry. Several million hectares of saturated soil can be potentially exploited from rainfed lowlands of eastern India.

The India Exclusive Economic Zone (EEZ) of the marine sector, of about 2 million sq.km., has high potential harvestable yields. Only about 20% of the cultivated area is covered under tractor cultivation, great scope of making quantism quantum jump in production and productivity exists, changing consumption patterns offer scope for diversified agriculture. Trade liberalization provides opportunities to reach for outside markets, which were not accessible otherwise. Great gains are possible through agro-based industrialization and overall rural development.

The cultivators and farmers of India have a rich heritage of traditional agricultural wisdom on crop and livestock husbandry and fisheries; based on the principles of the conservation of natural resources and environment. India has an extensive network of extension channels for dissemination of technologies generated by research institutes. India has a strong manufacturing capability and a large network of about 20,000 manufacturers and nearly one million village artisans for indigeneous production of all types of agricultural machinery. Therefore, inspite of several disadvantages, we have a great chance to leap forward, if we enforce the right policy initiatives and implement them.

Biotechnology and India Agriculture

We need to recongnise that, just as the food reequirements of today's population of nearly 6 billion people could not have been met by the technologies of the 1940's, we cannot assume that current practices will feed the population of 8 billion expected by 2020. New approaches are needed in addition to the continued improvement of existing methods of crop and animal husbandry and food processing.

The strategic integration of biotechnology tools into India agricu-ltural systems can revolutionise Indian farming and usher in a new era. Genetic engineering is clearly the most revolutionary tool to impact agricultural research since the discovery of genetics by Mendel. Prior to genetic engineering, the exchange of DNA material was possible only between individual organisms of the same species. With the advent of genetic engineering in 1972, scientists have been able to identify specific genes associated with desirable traits in one organism. For example, a gene from bacteria, virus or animals may be transferred into plants to produce genetically modified plants having changed characteristics. This method therefore allows mixing of the genetic material from species that cannot otherwise breed naturally.

Genetic engineering can be vital for an agrarian country like India. It can help in minimising the crop damage through disease and pest resistant varieties, reducing the use of chemicals, enhancing stress tolerance in crop plants, thus permitting productive farming on unproductive lands, etc. One could even extend the growing season of crops and minimise losses due to environmental factors on the one hand and increase the shelf life of fruits and vegetables, on the other, thus minimising losses due to food spoilage.

We can expand the market vista and improve food quality. Biotechnology can also produce plants that possess healthy fats and oils, possess increased nutritive value, and create a whole range of higher value feeds. Biotechnology has even the power of producing biodegradable plastics, edible vaccines, etc. It is only through the blending of the 'gene revolution' with our experience in the 'geen revolution' that we can reach our goal of 'evergreen revolution' and also 'nutritional revolution' The advantage of the gene revolution is that it is relatively scale neutral, benefiting big and small farmers alike, It is also environment friendly.

Thus, it can be of great help to the smallest farmer with limited resources in increasing farm productivity through the availability of improved but powerful seed. It can also reduce a farmer's dependency on chemical inputs such as pesticides and fertilisers. Modern biotechnology offers unlimited opportunity for enhancing genetic potential of crops and other commodities, management of biotic and abiotic stresses, bio-remediation and organic recycling. India is one of the main centres of agricultural biodiversity and its gene richness can greatly complement the developments in modern biotechnology. Likewise, new developments in GIS, remote sensing, and crop modelling, provide new opportunities for integrated mana-gement of natural resources.

The revolution in informatics provides opportunities for sharing latest information for research and planning in a highly organised and efficient manner. With the globalisation of economy, the opportunities for value addition and post harvest management are also immense. With clever blending of technologies, the India farmer can usher in new era of confidence and performance.

Sir Francis Bacon had once said:

> *'It would be an unsound fancy to expect that things which have never yet been done, can be done except by methods which have never been tried'*

In the new context, we have to review the new role of biotechnology. Also, it is in this context that we have to view the rapid advances that modern biotechnology is making, including the area of transgenics. Let us focus on some of the developments in this area, since considerable interest and controversy has been created around the world today in this exciting area.

Movement in Transgenics

The advent of genetic modification over the last two decades has enabled plant breeders to develop new varieties of crops at a faster rate than was possible using traditional methods, with huge potential for further beneficial development. Tobacco was the first plant to be genetically transformed\, in 1983, with cereals beginning in 1990, but it is only recently that products such as GM soya have reached the market place in Europe, and, for a variety of reasons, given rise to concern & controversy.

The concern must be addressed seriously if society is to exploit the new technologies appropriately. We will address this issue rather extensively a little later. Whereas in India, we have been discussing and debating the issues, the rest of the world is galloping ahead.

Table: *Ag biotech product sales*

$ Millions	***Sales***			***Annual Growth Rate***	
	1994	1997	2002	1994-97	1997-02
Transgenic seeds and plants	20	405	2,030	173%	38%
Animal growth horomes	80	225	405	31	12
Biopesticides	35	65	110	23	11
Other	107	180	340	19	14
Total U.S. Sales	$242	$875	$2,885	53%	27%

Development, field testing, and commercialisation of transgenic crops, is now recongnized by the international scientific community to be an essential strategy for food security. Fortyfive countries, including India, have conducted transgenic crop field trials up to 1997. The activity in the developing countries has been shown in Table.

Table: *Releases of transgenic plants in developing countries (by species and introduced trait)*

Species		***Introduced Trait***				
	Total Fied	***Herbicide Resistance***	***Insect Resistance***	***Virus Resistance***	***Product quality***	***Others releases***
Maize	46	29	16	1	3	5
Soya bean	27	25	—	1	1	—
Cotton	24	16	15	—	—	—
Tomato	19	—	2	1	16	—
Potato	13	—	1	6	—	6
Subtotal	129	70	34	9	20	11
Other species	30					
Total	159					

The emphasis has been mainly on traits for herbicide, insect, virus & fungal resistance, and on product quality with a limited number of genes. Out of a total 159 releases. Argentina leads the way with 43, followed by chile, Mexico, Puerto Rico and the Republic of South Africa, with 17-20 each. India is far behind with just a few releases. There are fears relating to transgenic plants in society. These need to be addressed, not by brushing them aside, but by careful scientific analysis and educational programmes.

All biosafety protocols must be followed rigidly. Let us deal with this issue more exhaustively now.

Public Concerns

The potential of biotechnology as a method to enhance agriculture productivity in the future has been accepted globally. However, because of its revolutionary nature, risk and uncertainty may be created by the process of genetic engineering and by the resulting genetically modified products. This may result from the introduction of a new unrelated DNA sequence into a recipient organism. The introduced DNA may have some unexpected effects on the cellular processes of the recipient organism. In addition to this, introduction of antibiotic resistance genes, as selection markers, pose serious implications in public health especially in genetically modified plants that are directly used for food purposes. Risks are also associated with the genetically modified plants that are released into the environment. The nature of interactions with other organisms of the natural ecosystems cannot be anticipated without proper scientific testing. For example, modified plants with enhanced resistance to pests or disease threaten to transfer resistance to the wild relatives, which will have implications for biodiversity and ecosystem integrity. These, and many more, doubts plague the minds of common people.

Therefore, it is very clear that if the biotechnology potential has to be completely realized, it has to be done in an extremely responsible manner. At each step, proper testing and scientific data have to provided. Not only this, proper regulatory and policy mechanisms have to be put into place. It is therefore, absolutely essential, that conscious efforts be made especially by the developing countries to initiate well-defined programmes for the development and regulations of genetically modified plants.

To make this possible and to benefically, ethically and sustainably reap the benefits of biotechnology, there is a pressing need for scientists, researchers, policy makers, NGOS, progressive farmers, industrialists, and representatives of the government, to come together on a common forum in order to discuss the common concerns and find solutions to them.

Will modified plants transfer their introduced genes into wild relatives growing nearby?

Will modified plants that produce new compounds disrupt the 'balance of nature' in some way?

Could the planting of a restricted number of cultivators leave crop plants more susceptible to disease?

Could the planting of a restricted number of cultivators lead to a reduction in biodiversity (of crop plants, weed species, insects and microflora in the fields in question)?

Will genetically modified plants be able to avoid the factors that regulate natural populations and thereby change the usual ' balance' between populations?

Similarly, there are concerns about the social and economic effects. For example, some of these concerns are:

- How will the structure of farming (particularly in developing countries) be affected by biotechnology?
- How will patent laws affect traditional breeders 'rights (e.g. the right to save seed from one year to the next)?
- Will plant breeding be left increasingly in hands of a few companies, and if so, what effects might this have?
- Will some countries be 'plundered' for their genetic resources?
- We also have a set of ethical and moral issues that need addressing. In particular,
- The consumer has the right. What will consumers be told about the new food products?
- Is it acceptable to 'interfere' with nature through genetic engineering?
- Do we have the right not to use all means available to improve crop plants, especially when so many people are under or malnourished?

Then there are several regulatory issues. Among these, the three key issues are:

- Do current regulations give sufficient protection to farmers, consumers, those who have invested in, and those engaged in research?
- Is there sufficient international legislation to ensure environmental protection?
- Do current regulations compromise the competitiveness of biotechnology companies by being excessively restrictive?

The Royal society of London had appointed a group of experts to examine various aspects including the scientific evidence concerning the risk of transfer of genes from Genetically Modified (GM) crop plants to wild species and non-GM crops, the uptake of genes from GM food by the digestive system, and the current state of the regulatory system. The experts concluded that the chances of gene transfer happening are slight, provided the regulatory processes are followed, but that this must be kept under consideration. The experts also concluded that the uptake of genes via the food chain is not a new issue because genes (i.e. DNA) are normal constituents of the human diet. Many products from GM Plants, such as sugar prepared from GM tomato paste, are so similar that they are regarded as 'substantially equivalent' others, for example flour from GM soya, may contain a new gene or its product, although many of the purification processes involved in food production will destroy and DNA present in the raw material. Some GM foods have been produced using an antibiotic resistance 'market' gene, which is a laboratory device designed to identify genet-ically transformed plants. The Royal Society has shown concern about the use of such genes in food products and it suggested that any further increase in the use of such markers in the human or animal food chain would be undersirable.

Gm crop plants have been produced to improve insect tolerance and virus resistance, and to include herbicide tolerance, so that transferred to non-target species of plants, and that the development of resistance by target pests is minimised, the regulatory authorities must be assured that any negative effects would be no greater than those resulting from conventional procedures; and that any long term effects on the environment and ecology would be closely monitored, with statutory restrictions in place to control marketing; and also that best practice advice was adopted by growers. The reliance on a case by case approach to the legislation, may result in lack of analysis of the overall impact of the technology on agriculture and the environment, and of the long term effects of gm. Although mechanisms are already in place to regulate many individual aspects of GM technology, there is no means for looking at GM technology as a whole. As independent, overarching, regulatory body is needed to span departmental responsibilities, monitor the enforcement of existing or future regulations, and strengthen the guidelines to growers of GM crops, such as those specifying the isolation distances between Gm and non-GM crops. The body should also review and monitor the membership of advisory committees and regulatory bodies.

The Indian Effort in R & D

Indian has set up a large network of research institutions and agricultural universities located in different regions and states of the county.

Biotechnological Applications Relevance to The Indian Farmer

Increased food production must keep pace with the country's population increase. Not only the quantity of food made available should improve. To keep pace with the present population growth and consumption pattern, India's food requirement has been estimated to cross 225 million tons by 2005 AD. This would mean an annual agricultural growth rate of 6.7 percent, a daunting task considering the rapidly shrinking resource-base and fast declining input-use efficiency in major cropping systems. Not withstanding the impressive gain in the agricultural production, the vast agricultural potential still remains highly under-utilized. There are serious gaps both in farm yield realization and technology transfer, as the national average yield of most crop varieties of most crop varieties is low.

Not many superior varieties is mot pulses, many oilseeds and vegetable crops have been evolved in recent past. In case of vegetables the varieties which were imported or developed many decades ago are still being grown by the farmers. Their yields are low and their inherent ability to respond to improve production technology is also limited. Farmers have accepted bybrids in my vegetable crops such as cabbage, tomato and capsicum.

Research and Technology Transfer

The Central and State governments continued to invest heavilyu in research, hence the Indian public sector research system has increased in size in real terms but by mid 90s levelled off. The number of hybrids and varieties developed and released by ICAR Crop Improvement Projects and State Agricultural Universities in major and minor crops are enormous and have been well documented.

Private research to develop superior hybrids was initiated by few seed companies in mid 60s. By 1987-88, the number increased to over 12 and in 1997-98 it is estimated at 34. With the governments recognition of provate sector's performance capabilities and with "New Seed Policy" announced in 1988, the number of private companies involved in crop improvement programs have been steadily increasing (Table).

Number of Private Companies Engaged in Crop improvement Program (1987-1998)

Table: *Number of Companies with Crop Improvement Programs*

Crop	*1987-88*	*1993-94*	*1997-98*
Pearl Millet	12	23	25
Sorghum	10	12	15
maize	06	15	20
Sunflower	10	30	20
Cotton	09	21	34
Hibrid Rice	-	4	6

For want of financial, infrastructural and trained human resources, most private companies specialize in breeding superior hybrids in few crops characterized by low volume, high value and amenable to hybrid technology. In addition to the ICAR and Agriculture University research system, about 13 private sector seed companies are actively involved in vegetable R&D. Some of them have joint ventures with multinational or companies specializing in vegetable crop research and utilize research facilities available with their overseas partners. Few have created facilities for field testing imported and public bred varieties. The role of private sector in vegetable research is primarily in the development of F hybrids and few open pollinated varieties.

Table: *Hybrid Vegetable and Melon seeds of Private R&D available in the market (1996-97)*

Crop	*No. of Companies*	*No. of hybrid Varieties*
Tomatoes	11	45
Brinjal	10	38
Chilli	8	27
Bhendi	13	25
Cabbage	11	40
Cauliflower	8	21
Ridgegourd	6	8
Watermelon	5	13
Muskmelon	2	7

Number of private/public hybrids in major crops in market during 1995 are compared with the number in 1997 in table.

Table: *Number of private and public bred hybrids in major crops in the market during 1995 and 1997.*

Number of Hybrids in Market

	1995		*1997*		
Crop	***Private***	***Public***	***Private***	***Public***	***No of Private Hybrids grown on 2% plus area***
Pearl Millet	50	04	60	06	14
Sorghum	22	04	41	05	6
Maize	57	03	67	03	12
Sunflower	47	05	35	06	10
Cotton	73	15	150	15	19
Hybrid Rice	4	4	8	4	-

In pearl millet, sorghum, maize and cotton number of hybrids introduced by private, companies has been steadily increasing. Many hybrids especially in paddy and cotton have regional adaptability. The number of hybrids which are accepted by the farmers and may be grown on 2 percent or more area is much lesser, so also the number evaluated in coordinated project trails or identified for release. Many hybrids recently supplied by the private sector embodies new technology which is either imported or developed by private companies. Private R&D's real investment in research has quadrupled between 1986 and 1997. Subsidiaries or joint ventures with multinational companies account for 30 percent of all private seed industry research in 1997. Both domestic companies and multinationals companies have started research programs on rice and mustard. Recently, Monsanto has also focused on hybrid wheat. Few private sector companies have also initiated joint ventures (e.g. Mahyco-Monsanto Biotech, ProAgro-Agro-Evo) to transfer specific genes to their product lines to provide effective protection against pests. Some of them have taken initiative to treat their seed with hightech chemicals to provide protection against sucking pests (aphids, jassids, thrips) during seedling and grand growth crop stages and diseases (downy mildew in bajra) during flowering.

Commercial Seed Production

Majority of certified/commercial seed is produced under contract with trained seed multipliers in suitable agro-climatic conditions.

Both public and private companies make contact with progressive farmers or trained seed multipliers and negotiate terms and conditions of production. Individual seed multipliers have small areas to contract and this greatly increases the cost of production and administration. To overcome this situation, contracts are arranged through local organizers, who in turn contract individual farmers, consolidate individual fields intoblocks and supply to the company the ultimate product.

Companies generallyprepare a technical package for seed multiplication specifying the kind of soil needed, isolation distances required, staggered plantion in case of a hybrid if needed, distances among rows and plants, depth of planting, time of nutrient and spray applications, frequency and time of irrigation, mode of harvest especially in case of hybrid seed etc. They prefer to contract between 300-400 acres in a village and designate such an area as a "seed village". They create seed processing facilities and place technically trained staff to guide seed multipliers, monitor actual production and supervise seed processing. Generally one such unit serves a cluster of 5-10 seed villages Only seed lots meeting genetic Purity, Physical Purity, germination and disease standards prescribed by the government of India, are packed and commercially sold. User-farmers have become very consc-ious not only of genetic purity and germination, but also of physical attributes of seeds. They prefer bold, shining and plump seed, unaffected by mould The companies therefore, organize seed production in areas with appropriate agro-climatic conditions and no rains in the post-flowering period of seed crop.

They also ensure proper cleaning to remove undersized, premature, broken seeds or inert matter on cleaner-cumgraders, gravity table, de-stoners and thus improve physical attributes of seed lots. Seeds are treated with recommended pesticides to prevent them against stored grain and soil pests. To add value to their seed, many companies now treat them with specific chemicals to provide protection agains sucking pests during seedling stages or specific diseases such as downys mildew for 40-45 days after emergence.

The seed production of most hybrid crops is generally undertaken in Western and Southern parts of the country because of congenial agro-climatic conditions and possibility of taking two seed crops in a year if necessary. The prominent states are Gujarat, Maharashtra, Andhra Pradesh, Karnataka and Tamil Nadu. In contrast, the Northern states from Punjab to West Bengal are dominated by rice, wheat, and legume crops which are less profitable. Consequently there is a major difference in the cadre of seed industry between the North and West/South of India. Technology of seed production is moving at a rapid pace. Hence the companies have to ensure that they pass on the technology to the contract seed multipliers to harvest high yields of good quality seeds. Most public bred hybrids and variety seeds are certified while the proprietary hybrid seeds are sold under truthful label. All seeds meet minimum prescribed standards but for proprietary hybrids quality monitoring is through their in-house system.

Private Hybrid's Share of Seed Market

Private sector has played a major role in supplying quality seeds of both private and/or public hybrids. It is interesting to note that 69% of the pearl millet seeds produced during 1996-97 season was of private hybrids. Other crops where private research plays key role in seed supply are sudangrass sorghym (100%), sunflower (98%), Maize (90%), sorghum (52%) and cotton (50%). The contribution of private research in value terms is also steadily increasing. The share of research hybrids to total turnover in these crops was about 70 percent in 1996-97 as compared with 46 percent in 1990-91. Private hybrids are successfully meeting the regional needs of early maturity, tolerance to pests, unpredictable agro-climatic condition and soil variations.

Government Intervention: Legislation and Policy Issue

At the highest level, government policy is to reduce controls and to make the economic system more transparent in practice however, there is potential proliferation of legislation affecting seed sector. Given the current climate of economic development there is an urgent need for positive relationship between the government and the private sector. In order to fulfil farmer's needs and GOI food production targets set for the year 2005, the existing misapprehensions should disappear to ensure private sector investment in infrastructural improvement and expansion to increase quality seed production.

Presently Indian seed industry is governed by the following Acts:

- Seed Act 1966
- Seed Rules 1968
- Seed Control Order 1983
- Essential Commodities Act 1955
- Package Commodities order 1975

- Standrads of Weights and Measures Act 1976
- Consumer Protection Act 1986.

Export Regulations and Quarantine :

- Plants, Fruits and Seeds (Regulation of import into India) Order 1989
- Urban Land Ceiling Act 1976
- State Acts of land acquisition/land use
- State Acts for the control/movement of crops and seeds.

The policy framework hitherto has been regulatory with excessive legislation. Above plethora of legislation covering the seed sector is out of date and restrictive rather than progressive. The multiplicity of Acts, rules and administrative orders have resulted in over regulation of the nascent industry.

Future Trends

During past four decades spectacular progress has been achieved in agricultural production in the country. Compared to 51 million tons of food grain production in 1950-51, India realized a food grain production of over 196 million tons during 1997-98, resulting not only in self-sufficiency in food but also in a buffer stock of over 30 million tons. Food grain production is expected to cross 200 million tons in current year (1998-99). In the oilseed sector, which is primarily production was coupled with important policy planning and research and technology development transformation.

The steady increase in productivity has been accomplished by exploiting hybrid vigour in maize, sorghum, pearl millet, sunflower, cotton, castor and improved genetic potential of high yielding varieties in wheat, paddy, many legumes and oilseed crops. The superiority was for high inherent harvest index, inbuilt resistance to major disease-pest complex, ability to withstand environmental stress conditions, adjusting growth cycle to better suit the available growing conditions, generally greater physiological efficiency and/or improved quality. Nineties focused on exploiting heterosis in rice, rapeseed-mustard and cajanas for enhanced productivity, sustainability and utilizing the vast agricultural potential which still remains highly under-realized through bridging the serious gaps in inherent yield potential and ensuring a rapid technology transfer to increase national average yields leading to increased produ-ctivity. The Indian seed industry shows a clear pattern of development that also reflects in government policies. The decade, after the announcement of New Seed Policy of 1988, witnessed entry of major multinational companies and foreign genetic material to the Indian seed market. Nineties also saw proliferation of domestic seed companies into more dynamic prevailing economic climate. Although public sector dominance continues in open pollinated variety seeds, the private sector is steadily increasing its market share of hybrid seeds. Commercially the seed industry is going through a period of accelerated development with activities now including in-house research, adoption of agro-biotechnology and genetic engineering techniques.

Till such time that the population stabilizes, the demand for food would continue to increase. To keep pace with the present rate of population growth and consumption pattern, India's food requirement has been estimated to cross 225 million tons mark by 2005. The certified/quality seed distribution has considerably increased from a mere 18,300 tons in 1952-53 to over one million tons in 1997-98. The quality seed requirements are estimated to increase to 1.8 million tons by 2005 AD. If the population growth continues at current rate, the country could require 300 million tons of food to feed an estimated 1.4 billion people by the year 2025. This will only be possible with an average agricultural growth rate of about 5 percent annually. India, therefore, need to continue the green revolution through introduction of value added superior genetic material possessing inherent high yield potential even under environmental stress conditions and genetically engineered or chemically induced seed treated tolerance to major disease-pest complex. Then only the income levels of small and marginal farmers who dominate Indian agriculture today will rise. Accomplishing such targets of production will not be an easy task considering the fast declining input use efficiency for major cropping systems and rapidly shrinking resources base.

Future agricultural development would be guided not only by the compulsions of improving food and nutritional security but also major concerns for environmental protection, sustainablity and profitability. Following GATT agreement and the liberalization process, globalization of market would call for competitiveness and efficiency of agricultural production. Projected population migration from rural to urban areas will also demand a significant shift in the pattern of the food basket. Developments in biotechnology, molecular biology including genetic engineering and new cutting edge technologies in recent years will provide new potential for accelerated growth. Used in consort with conventional approach and harnessed judiciously, these new break throughs would help increase agricultural productivity and sustainablity. To meet the future challenges of increased productivity-a daunting national need, the problem oriented, demand driven, value added approach of integrated research and development involving all reasonable players as partners would offer practical solution.

This would necessitate an active cooperation between the public and private solution. This would necessitate an active cooperation between the public and private sector agricultural research, massive private invest-ments in R&D, global exchange of germplasm and technology and necessary changes in government policies to create a congenial environment to encourage entry and active participation of technology rich foreign private companies in Indian seed market. Investments and technology would needed to breed hybrids and varieties with high inherent genetic yield potential with value added traits, incorporated through genetic engineering to improve nutrition, provide protection against major pest-disease complex, inbuilt drought tolerance to ensure increased productivity even for arid and semiarid zones of the country. Biotechnology applications would not only minimize loss of biodiversity but will also expedite introduction of high yielding traits in desired crops of major economic importance.

ICAR is currently preparing a new perspective plan to upgrade farm research and man power development systems in the country. They are being reoriented to ensure sustainability and globalization of agriculture with emphasis on diversification, value addition and to encourage exports. This is opportune time for bold initiatives and policy decisions in support of private R&D which is steadily becoming stronger. The number of proprietary hybrids is increasing rapidly. Hybrids in rice are being commercialized. Hybrids in rapeseed-mustard and cajanas are on threshold of commercialization. Research efforts to evolve hybrids in wheat are on. Innovations in genetic engineering are being incorporated in Indian products; in cotton to provide resistance to Heliothis bollworm; in brinjal, cauliflower and cabbage among vegetables for black moth. Genetically engineered products will probably enter commercial market in early next decade. The present value of formal seed market estimated at Rs. 20,190 crore, i.e. US$ 480 million with private sector contributing approximately 72 percent, is likely to double in next 3-4 years. There is lot to be gained from these synergies that remain untapped between government laboratories, industrys and academia. It is an appropriate time for an effective partnership between ICAR projects and institutes, state agricultural universities and private R&D to evolve. Time is ripe for a second green revolution which would not only concentrate on increased productivity but also on value added traits to reduce cultivation cost, pollution dangers and improve quality of products. Rapid developments in biotechnology, particularly in genetic competition.

Indian seed industry is ready to face the challenge. Government is also cognizant of need to remove bureaucratic hurdles to its expansion. It is most opportune time for government to take bold decisions in favour of private sector development. It needs to start recognizing and rewarding innovations and encourage domestic companies to invest greater resources in research. Amended Indian Patents Act Of 1910 includes biotechnology and genetically engineered product patents and has been Passed by Rajya Sabha (Upper House). Government should expedite implementation of Plant Variety and Farmer's Rights Protection Bill. These steps would send right signals to the international of Plant Variety and Farmer's Rights Protection Bill.

These steps would send right signals to the international seed market and would encourage massive investments both from multinational and domestic seed companies to set up high-tech R&D, seed production and conditioning facilities to produce seed meeting international standards, for both domestic market and for export. Seed industry needs to become technology driven and innovation led. This transition will have to be facilitated through innovating strategies where both government and industry are partners. The private sector's R&D emphasis on hybrids would also then include open pollinated varieties in major self-pollinated crops where hybrid systems are not yet identified. As the business expands, large companies would create a full line of products to meet their dealer's demands. Since most of this technology will be high value proprietary products, many farmers are likely to become contract growers. Their risk will be limited, and profits although smaller will be guaranteed. New technology thus holds the potential to make farming more secure and profitable.

The guiding philosophy of ICAR research for next millennium's hi-tech environment should be to complement private sector R&D efforts. ICAR's "research product mix" should reflect priorities and targets that continuously evolve in response to changing internal and external research environment. Its policies should consider a shift in emphasis to upstream research producing mainly intermediate outputs, i.e. inputs to further research, which in turn will yield better products. In past 4-5 years over 19 billion US dollars have been committed to mergers, acquisitions and technology alliances in the crop genetic industry world wide.

There has also been incredible growth of big com-pany R&D budgets in this area. The agriculture in the developing world is becoming like any other industry and is being dominated by a hand full of big technology developers. Until a decade ago hundreds of competing companies except for one or two big ones accounted for nearly half of total seed market share. Now the technology is a hand full of large companies developing the technology and controlling much of the industry and fewer smaller players. This worldwide changes would have a major impact on the pattern of Indian seed industry growth in the coming century.

Seed industry worldwide is also going through a rapid technology revolution with value added output traits at its centre, driven by consumer and commercial end users. It involves all aspects of agri-biotechnology and crop production. Emphasis is on proprietary genes in value added crops for the feed, food, chemical and pharmaceutical industries at the centre of new seed industry. This change to value added agriculture will also have major impact on restructuring of Indian seed sector, hence seed business in coming decades. It will become more difficult for seed companies to assess new technology. In the changing seed industry scenario, seed companies in India will continue small-scale plant breeding, product testing and maintaining a strong, localized seed distribution network with cordial customer relationship. Using this strength they can get involved in alliances and agreements that may ensure access to the future technology. The future trend is quite clear from Monsanto's equity acquisition in Mahyco; Mahyco-Monsanto Biotech Ltd.; Mons-anto acquiring Cargill and buying equity in E.I.D.Parry's seed business; Agr-Evo acquiring Pro-agro; Semenis buying 40 percent additional equity stock in Nath Sluis and so on. Hence the Indian seed companies will continue to play a basic and important role.

Like present day contract grower farmers, small seed companies of all sizes will incre-asing become contract producers and developers for the bigh technology rich companies. Because of rapidly changing new technology the seed industry world over is going through a radical change with a whole new future of rapid growth ahead of it.

Indian seed sector would have no option but to follow that path with active government support and huge private investments wherever needed. A large cadre of highly skilled seed multipliers exist. Companies both in public and private sector have developed systems for training technical man power to monitor expanded infrastructure.

As their R&D gains strength, they will develop new hybrids and varieties suitable not only for Indian market but also for the agro-climatic conditions prevailing in the export market. India's low cost skills at all levels of seed industry from research through seed production coupled with range of environments suitable for production of most crop seeds and rapid flow of technology offers excellent base for competitive supplies in the international market. Indian seed industry has reached a stage of maturity and is in a position to shape itself to compete at national and international levels. This may however, be recognized that private sector for some more time will continue to concentrate on more lucrative sector of the market, with a strong emphasis on hybrids and the more specialized low volume crops, e.g. vegetables. Fundamental strength and potential of Indian seed market cannot be doubted. The country has a very large population and a strong commitment to self-sufficiency in food production. This can only be achieved by increasing productivity per unit area.

Indian economy is opening up the opportunities and challenges offered by the globalization of agriculture, particularly in the seed sector. Given that the prevailing uncertainites about legislative changes which currently retard seed industry growth are favourably clarified, the present trend of industry's accelerated growth in domestic market, value addition to their products, custom production and export will continue. Indian seed industry is on the threshold of becoming a global player. Beginning next century, Indian private sector can look forward to playing a key role in the development of Indian seed industry and take its rightful place in the global market.

8

Biotechnology: Fungicides and Insecticides

Our doctors are surprised with the rising rates of obesity, diabetes, hypertension, heart disease and cancer. Most obese women are not aware that they are diabetics till visit the doctor for some illness or surgery. Hypertension too, being a gradual rise of blood pressure, most people are not aware until they feel dizzy, and some end up with heart attacks and strokes.

Knowing about nutrition, healthy living and longevity today has become so specialized that most people tend to get confused not knowing whom or what to follow. There is a diet revolution today and the more you read about cholesterol, saturated fats, butter, margarine, carbohydrates, etc. the more you get confused. The answer is, read every bit you could gather through the Internet, magazines, newspapers, and nutritionists.

Americans have given us their heritage by the large imports of wheat flour for our traditional foods for over a century. The process that wheat goes through to obtain that fine white flour we import from the States are as shocking as the process margarine goes through to make it appear like butter. Even before they are planted in the ground, wheat seeds receive an application of fungicides and insecticides. Fungicides are used to control diseases of seeds and seedlings; insecticides are used to control insect pests, killing them as they feed on the seed. Some of the chemicals used in commercial wheat crops are disulfo-ton, methyl parathion, chlorpyrifos, dimethoate, diamba and glyphos-phate. These chemicals increase the toxic load in our bodies, leading to increased susceptibility to neurotoxic diseases as well as conditions like cancer. Many of these pesticides function as xenooestrogens, foreign oestrogen that can reap havoc with our hormone balance, states Jen Allbritton, a certified nutritionist in the States. He further states that the researchers speculate these oestrogen-mimicking chemicals are one of the contributing factors to boys and girls entering puberty at earlier and earlier ages. They have been linked to abnorm-alities and hormone related cancers including fibrocystic breast disease, breast cancer and endometriosis.

Farmers also apply hormone like substances or plant growth regulators that affect wheat characteristics, such as time of germination and strength of stalk. Cycocel is a synthetic hormone that is commonly applied to wheat. During the process of storage of wheat, the collecting bins are sprayed with insecticide, inside and out. More chemicals are added while the bin is filled. These so called "protectants" are then added to the upper surface of the grain as well as four inches deep into the grain to protect against damage from moths and other insects entering from the top of the bin. One could imagine how much poisonous linoleic acid gets into the system by regular filmgoers, especially the kids who consume big loads of popcorn during the intervals.

Two slices of whole-meal bread with lentil curry and pol sambol, would be an excellent morning choice. Pol sambol is nutritious, scraped coconut in addition to its nutrients, antioxidants, has lauric and capric acid having anti-microbial properties. What better way to start the morning with food that kills viruses, parasites, bacteria and fungi? The chillie powder in the sambol has its vitamin C, and increases the general metabolism, also accelerates the functions of the glands. Added lime juice gives the fruity taste and vitamin C. The Maldive fish adds the omega 3 oil, good for heart health. Foods are now rated on a scale called glycaemic index, or simply 'GI'. The carbohydrates that are high in GI are less healthy. White bread (GI of 70) is used as a reference sometimes instead of glucose. The foods that take longer to be absorbed are called 'low GI (GI less than 55). The carbohydrates that are quickly digested and absorbed are called 'high GI' (GI greater than 60).

Lentils are nutritious, and have a GI of less than 30. It's a good protein for the vegetarians and diabetics. Mun-ata (green gram), lunu miris, and scraped coconuts are excellent foods as a stomach filling breakfast. Mun-ata also has a GI less than 30. Other low GI foods are Soya beans (GI 14); Peanuts (GI 15); Red lentils (GI 18); Kidney beans (GI 27) Apples (GI 38); Spaghetti (GI 41); Orange juice (GI 46); Raw carrots (GI 49); Cooked carrots (GI85); Baked beans (GI48), and so on.

Some studies have shown that a diet based on low GI foods can lower the blood cholesterol, lower risk of heart disease. It is clear that if one craves for starchy traditional foods, home-pounded rice flour would be a better choice than the imported white wheat flour from the States.

Save our Genetic Resource

I was surprised to see at Aricha Ghat, Bangladesh that vegetable and fruit sellers were shouting, "Buy Desal (country or wild variety) of fruits!" These are more expensive than imported modified varieties. But taste better and contain more vitamin and minerals but smaller in size. Local varities iof rice or pulses, spices in village market are more expensive. But people have realize that traditional varities are better. But the big NGOs are selling imported manipulated seeds promising high yields which need chemical fertilizers and pesticides.

Farmers of Indian Sub-Contient used to plant traditional seeds but now after introducing high yield varities there is always an artifical seed crisis. Whenever there is a shortage of a commodity, black marketers lose no time in starting their profit making scheme, and this time too they have surfaced in the area. Our correspondent from Kurigram writes that against the BADC's fixed rate of Tk. 180 for a 10-kg bag of boro seed, black marketers are selling it for about Tk. 230. While this has been going on quite in the open in the district, BADC authorities, as usual, have been denying any such corruption. We are surprised to further learn that some officials have even denied any seed crisis at the sales centres claiming that sufficient quantity of seeds would reach there within hours.:

Seed Crisis hits Boro (HYV Rice) Farming

Crisis of seeds is hampering cultivation of Boro paddy in the district in the current season. Farmers of the district are facing problems in procuring Boro seeds from Bangaladesh Agriculture Development Corporation (BADC). Despite remaining standing in the queue for long while the farmers are returning home without seeds from BADC sales centres. This correspondent visited BADC's seed sales centre in Kurigram recently and saw farmers standing in a long queue to procure seeds. A 10-kg bag of Boro seed is being sold between

Tk 205 and Tk 230 in the market against the government rate of Tk 180, they alleged. Besides, some dealers are selling seeds in the black market at exorbitant price as BADC authorities have failed to provide the farmers with seeds in time. As a result, the genuine farmers are being deprived of seeds, they added. Hafizur Rahman, sub-assistant director (seed) of Kurigram BADC's sales centre, denied crisis of seeds at the centre and said that seeds would reach the godown within a few hours.

On the other hand, Mohiuddin Ahmed, Deputy Director of BADC, said that a total of 250 tonnes of seeds are required for cultivating 70,000 hectares of land in the nine upazilas of the district this season. But BADC does not have sufficient quantity of seeds at its godown. The BADC authorities have appointed 34 dealers for selling seeds to the growers in the nine upazilas of the district, but only three dealers have opened their sales centres, he added (The Independent November 25, 2003).

Narrowly focusing on increasing production-as the Green Revolution does-cannot alleviate hunger because it fails to alter the tightly concentrated distribution of economic power, especially access to land and purchasing power. Even the World Bank concluded in a major 1986 study of world hunger that a rapid increase in food production does not necessarily result in food security-that is, less hunger. Current hunger can only be alleviated by "redistributing pur-chasing power and resources toward those who are undernourished," the study said. In a nutshell-if the poor don't have the money to buy food, increased production is not going to help them.

Is it true that the best way to fight hunger, protect the environment and reduce poverty in Africa is by relying on Green Revolution crop varieties, and using more imported farm chemicals, plus genetic engineering and free trade? This is precisely what powerful institutions in Washington and elsewhere are prescribing for the continent, yet each of these elements could actually worsen, rather than improve conditions for Africa's poor majority (Institute of Food and Development Policy and the University of California at Berkeley, USA, 2002).

Agricultural Development Corporation (BADC) introduced modern irrigation equipment, chemical fertilisers, pesticides, and new varieties of seed that was developed by the international rice research institutes to increase food production in Bangladesh. Thus seeds developed in international research stations were made available to farmers for dry season (boro) crops in 1968 and wet season (aman) crops in 1970.

Farmers experience and advise were never consulted, while introducing modern irrigation system and chemical based agriculture, as they are regarded as"ignorant farmers". Introduction and implementation of "Seed-Fertiliser-Water" technology was financed by the World Bank, Asian Development Bank, IMF, industrial countries as credit and donation.

In Bangladesh traditional seeds are not available, and the poor farmers under bureaucratic and corrupt system are living near famine situation.

Acute Boro Seed Shortage in Nilphamari

Nov 28: An acute shortage of Boro seed is prevailing in the district, creating grave concern among the farmers. According to a source in the Agriculture Extension Department (AED) only 90 metric tons of Boro seeds were sanctioned by the BADC against the total demand of 27,328 metric tons in the district so far. The source revealed that this year the farmers of the district began to make their seedbeds in order to cultivate Boro paddy on 68.320 hectares of land. A total of 27,328 metric tons of Boro seeds are needed for it.

Every day the farmers are seen moving to and fro in collecting seeds.At present, the seeds of 10 kgs packet produced by the BADC are being sold at Tk 250 to Tk 300 in different hats and bazars in the district, whereas the rate of the Boro seed of 10 kgs packet produced by the BADC was fixed at Tk 170 for sale. When contacted, the Deputy Director of the district Agriculture Extension Department regarding the severe Boro seed crisis prevailing in the entire district, he admitted the reality(The Independent, November 29, 2004):

1. Serious Seeds Crisis
2. VALUE OF DIVERSITY
3. Botanical Garden needs care and protection.

Agriculture began between ten and fifteen thousand years ago through the efforts of hundreds of thousands of people on several continents, and in many different social and ecological situations. The agriculture they established, suited to their own needs, was developed over the course of several thousand years."

When human beings initially began to domesticate plants, they selected plants that had traits which ultimately would be of great advantage to people. For example, they collected seeds which were larger, matured at harvest time and were "non-shattering," meaning the seeds were not easily dispersed but clung to plants. The repeated selection and sowing of these seeds led to the development of plants more and more amenable to cultivation. But the multiplicity of places in which this process occurred ensured that a wide variety of plants, adapted to the particular requirements and situations of people all over the world, would flourish. Today, however, the international genetics-supply industry--made up of multinational chemical, pharmaceutical and food companies--is absorbing the world's small and medium-sized seed companies and strengthening its control over the world's crops.

In promoting their narrow line of seed varieties, the multinationals are rapidly pushing the industrialized countries toward an overall sameness in their food supply and threatening the world's genetic diversity.

At the same time, they are expanding their sales to farmers in the Third World, the origin of most of the world's indispensable plant varieties. By allowing this to occur, the world is inviting environmental disaster and widespread hunger.

Farmers who begin to rely on the high-yield seeds sold by the highly concentrated international genetics supply industry soon find themselves caught in a trap. Uniform crops are especially vulnerable to pests and disease, forcing growers to use pesticides, often sold by the same companies that provide seeds. Consequently, individual farmers become increasingly dependent on the genetics-supply industry. The integration of independent farmers into the worldwide corporate agribusiness system also has a negative effect on biodiversity. In expressing general alarm over the fate of the more traditional seed varieties

Fowler and Mooney stress that the extinction of a seed variety does not come simply at the time when there are no more seeds; rather extinction comes when the seeds' development process ceases to exist. As farmers become more dependent on the genetics supply industry, they lose the ability to nurture that development process.

Fowler and Mooney point out that approximately 97 percent of the varieties on a 1903 USDA vegetable list are now extinct. The consequ-ences of such uniformity were demonstrated by the 1970 corn blight, which destroyed over 15 percent of the U.S. crop. In the wake of the corn blight, a 1972 U.S. National Academy of Sciences report, Genetic Vulnerability of Major Crops, revealed that the United States was shockingly reliant on a handful of seed varieties for its major crops. Fowler and Mooney emphasize that the study concluded that U.S. agriculture was "impressively uniform genetically and impressively vulnerable." Tracing the origins of Latin America's coffee industry illustrates how dependent whole nations and regions of the world have become on a few crop varieties

The entire coffee industry of Latin America is based on the seven plants taken from Yemen a thousand years ago and then from one plant in Indonesia almost 300 years ago. Today, Fowler and Mooney report, there is little genetic diversity in coffee crops outside of Ethiopia. But the Ethiopian government, which believes the country has historically not received fair compensation for its genetic resources, is refusing to permit future collection of coffee resources from Ethiopia. "Unless Ethiopia relents," Fowler and Mooney write, "we will have an opportunity to see what happens to a narrowly-based crop like coffee without recourse to badly needed genetic resources.

The Struggle among Corporate Agribusiness Interests

Third World governments, the United Nations Food and Agriculture Organization (FAO) and the International Board for Plant Genetic Resources (IBPGR) to establish ground rules that will ensure the preservation and exchange of germplasm among all countries. Unfortunately, they report, international efforts at seed prese-rvation have favoured crops of primary interest to breeders in the industrialized countries. As founding staff members of the Rural Advancement Foundation International (RAFI), a nonprofit organiz-ation working for just and sustainable agriculture, both Fowler and Mooney have been participants in, and eyewitnesses to, many of those struggles. Based on their years of observations and careful research, the authors have sought to develop a constructive approach to the enormous political, economic and scientific problems that the questions of genetic diversity pose. They have concluded that following principles of genetic conservation are necessary to preserve a healthy and diversified seed culture:

- Decentralized, broad-based strategies of preservation.
- Multiple strategies must be employed, with as many groups as possible--including not just scientists, but farmers, fishing people and medicine makers--consulted about what should be saved and how.
- Agricultural diversity must be used, so that it continues to evolve and retain its value.

- And, efforts to save agricultural diversity must be interwoven with efforts to save the farm community.

"Diversity", like music or a dialect, is part of the community that produced it. It cannot exist for long without that community and the circumstances that gave rise to it. Saving farmers is a prerequisite of saving diversity.

Conversely, communities must save their agricultural diversity in order to retain their own options for development and self-reliance. Someone else's seeds imply someone else's needs." "We each have a special role to play in passing this gift [genetic diversity] on to the next generation," write Fowler and Mooney.

Throughout history, they show, it has been "amateurs"--people who love their seeds--more than scientists who have saved diversity.. If biodiversity is cared well in a country, the nation could be benefited by producing optimum harvest of its bioresources. Biodiversity then sponsors the sustenance of economic development of nation. So, we care biodiversity for two main causes for our existence in the world: a. We care for scientific/enviro-nmental reasons; and b. We care for commercial/trade reasons.

We care biodiversity for scientific/environmental reasons because it has great role in nature on the following headings: 1. Keeps the balance of biomass production in the biosphere; 2. Keeps the energy-flow in biotic form in balance in biosphere; 3. Keeps the soil fertility in balance in biosphere (eg. Keeps the proportionity of production of microbes in respective areas in the biosphere); and 4. Keeps the biotic-abiotic relations in balance (Ecosystem balance maintenance).

Heredity and the Gene

Knowledge of the principles of heredity is so basic to our fundamental understanding of the biological sciences that it is hard to believe that these principles were discovered only in the 1860s (and their importance was realized less than a century ago). However, a practical knowledge of the hereditary process came long before its mechanism was understood. Archeologists have discovered that as long as 7,000 years ago farmers in Central America were improving crops of corn by planting hybrid corn seeds that had developed preferred characteristics. Over 6,000 years ago, the Chinese learned how to develop superior strains of rice. An ancient Babylonian tablet shows a pedigree of a family of horses through five generations, with detailed information about height, length of the mane, and other traits, revealing that they had some knowledge that these traits were transmitted. Farmers and gardeners have continued to practice this type of selective breeding in both plants and animals. Each time an individual plant or animal appeared with a desired characteristic, it was bred again to produce more with similar traits. For example, at harvest time farmers would select heads of wheat that had the most or largest kernels and save them to use as seed the next year.

Science (or more generally, curiosity) received a large boost during the Renaissance. During the 16th and 17th centuries, the interest in the physical and natural sciences grew. This increased interest laid the foundation for science as we know it now. As gene technology bears on a large number of different disciplines (biochemistry, genetics, cell biology, engineering, etc.), a couple of key findings in different areas will be discussed here. It is interesting to note that some of the great discoveries that were made were not viewed to be very important at the time, just because the proper framework to understand the importance of the findings had not yet been developed. Also, it is important to realize that many of the seminal discoveries were made at the interface between two or more disciplines. The importance of this interdisciplinary insight and research toward developing new concepts and creating new ways of looking at problems persists to this day.

By the end of the 17th century, Hooke in England and van Leeuwenhoek in The Netherlands came up with microscopes. The microscope constructed by van Leeuwenhoek magnified 200 times, enough to see into "a new world" of little beasties that no one had ever seen. Hooke observed how cork was composed of a series of tiny cubicles, much like a honeycomb. He coined the word "cell" for such a cubicle, since it reminded him of the tiny cubicles used by monks in monasteries.

In the beginning of the 19th century, Brown, a Scottish botanist, found "opaque spots" in the cells of orchids. Upon further investigation, he found such spots in cells of all plants. He named the spot in the cell the nucleus, the latin name for "little nut". We will see later that the nucleus is the place where the genetic material is located and its expression is regulated. The current connotation of the word nucleus accurately represents the later realization that the nucleus is the "main office" of the cell, and has replaced the original meaning of the word.

It is striking that the discovery of the principles of heredity was not made by an eminent scientist, but rather by a monk doing experiments in the vegetable garden of a monastery. The monk,

Gregor Mendel, was intrigued by the multitude of shapes and colors of all living things, even occurring within a single species growing/ living at a single site. He used 34 different varieties of self-fertilizing peas for his experiments (self-fertilizing means that the egg under natural conditions is fertilized by the pollen from the same plant; in peas the flower remains closed during the fertilization period, thus preventing pollen from other plants fertilizing the eggs); he grew these varieties of peas for several pea generations.

He crossbred different kinds of peas by opening up a flower, cutting off the stamen, and putting pollen from another flower on the stigma. He tied a tiny bag over the flower to prevent any fertilization by other pollen. He worked on the problem for about eight years, and got results from 10,000 different plants. He selected seven characteristics to study (the color and the shape of the seed, the color of the flowers, the color and form of the pods, the position of the flowers, and the length of the stem), and system-atically focused on one or two parameters at a time. In his first exper-iment, he crossbred plants with round and wrinkled seeds.

The progeny only had round seeds, which was in contradiction with the theory current at that time, that the progeny contained a blend of the characteristics of the two parents! He grew plants from the progeny seed, wondering what the seeds of these plants would look like. This time, there were both wrinkled and round seeds, the round seeds outnumbering the wrinkled ones by a factor of three. In the next generation, the plants grown from wrinkled seeds only gave wrinkled seeds, whereas plants from round seeds once again produced a mix of round and wrinkled ones, in a ratio of three to one. He did similar experiments while looking at the other properties he had selected.

On the basis of the experiments he did, he formulated several rules: (1) heredity is determined by distinct elements contained in the two cells contributed by the parents of the organism, and these elements combine randomly; (2) each characteristic of an organism is determined separately from the others; and (3) when two different characteristics combine, one characteristic will dominate over the other (for his second and third rule he just had luck; it turned out to be not applicable in all cases). He decided to use single letters for each characteristic; a capital letter denoted the dominant factor (the one that determines the phenotype), and the same letter in lower case denoted the recessive one (the one that does not express itself in the phenotype, unless the dominant factor is absent). He realized that progeny combined factors from both parents, and that in the pollen or the egg only half of the information of the parent was present. This insight laid the foundation for modern genetics. He presented his findings and conclusions to a conference of local intellectuals. The response was a polite silence. Nevertheless, they invited him to publish his results in a scientific journal, which he did in 1866. His paper was totally neglected by the scientists of his time. He died 18 years later as a well-respected abbot, but without recognition for his pioneering work.

Mendel's pioneering work chronologically was paralleled by breakthroughs in biochemistry and cell anatomy. In 1869, a Swiss graduate student, Friedrich Miescher, set out to do a chemical analysis of a cell. He was having trouble breaking down all parts of the cell when he got the idea of "digesting" it with stomach enzymes. Cells broke down upon digestion, but they always left a residue from the nucleus. Miescher called the residue "nuclein". He combined pleasure with business angling for salmon in the Rhine river near his home. He had found that salmon sperm had very large nuclei, almost half of the cell itself, and Miescher was able to accumulate a good supply of his newly discovered nuclein.

It was fortunate that he did his experiments more than a century ago, because one would be hard-pressed to find any salmon in the polluted Rhine river now. One of Miescher's coworkers observed that nuclein was composed of carbon, oxygen, hydrogen, nitrogen and phosphorus.

Jars and jars of the white sticky stuff, the nuclein, stood on Miescher's laboratory shelves for years, but he did not realize that what he had isolated was a crude preparation of DNA, the genetic material that contained the molecular basis for Mendel's results. The chemical characterization of the nuclein was published by Miescher's friends after his death, but at that time it was just a curiosity, a piece of the puzzle that could not be fitted into its proper perspective.

In the middle of the 19th century, a young lad by the name of William Perkin was asked by his employer to find ways to synthetically make quinine, a natural drug used to treat malaria and other maladies. He went ahead, and mixed all kinds of chemicals. One mix, aniline and potassium dichromate resulted in a purple goo. Adding alcohol made a wonderful purple solution that did not cure malaria, but could be used as a dye. A quarter of a century later, a German scientist, Walter Flemming, stained some cells with Perkin's dye, to see how that would look under the microscope. Suddenly, the nucleus Brown had described half a century earlier became distinctly clear, since it absorbed the dye. Flemming called the mass of colored material in the nucleus chromatin. When he stained sections of growing tissue, he saw the cells in the process of division. He watched the chromatin bunch up into short, threadlike bundles, and he called these chromosomes. As he studied the bunching up process, he saw the chromosomes double in number. He watched the chromosomes begin to pull apart. Half of them moved to one end of the cell, the other half went to the other end. Because the chromosomes spread out like threads during this process, Flemming called the process mitosis ("thread" in Greek). Half the chromosomes moved into one of the new daughter cells that formed when the cell divided, and half went into the other new cell. They were exact duplicates. When Flemming wrote about mitosis and chromosomes in 1882, he did not connect it to inheritance. Mendel's laws had not yet been rediscovered.

Five years later, van Beneden, a Belgian scientist, discovered that every cell in a body has the same number of chromosomes, with the exception of the sperm or egg cell, which contains only half the "usual" number of chromosomes. He also found that different species have different numbers of chromosomes: a human has 46, a fruit fly has 8, and a crayfish has some 200. The giant sequoia can live with 22. Note that up to this point all important findings in this area had been made in Europe. Basic sciences in the US had not come of age yet.

Around the beginning of the 20th century, three scientists (De Vries in The Netherlands, Correns in Germany, and Tschermak in Austria) independently and virtually simultaneously performed plant breeding experiments that greatly resembled the work done by Mendel half a century earlier. All three only discovered Mendel's publication while doing a literature study in preparation of their publication. Around the same time, the name "gene" was coined to describe the hereditary substance, and "genetics" became the name of the rapidly emerging discipline of gene study and characterization.

Advances in the field of genetics were also fueled by the discoveries by Flemming, van Beneden, and others on the cellular/sub-cellular level. In the twenties and thirties of this century, chromosomes became the subject of intense study, and it is mainly the pioneering work of two groups (those of Morgan and Castle) that has contributed greatly to the further development of the concept of genes in relation to chromosomes. Both groups studied the fruitfly (Drosophila melanogaster), whose eight chromosomes are all readily visible under the light microscope. Six chromosomes could be divided up in three pairs of morphologically identical chromosomes, whereas the other two (named X) were identical in females while in males one of the two was replaced by a much smaller chromosome (Y).

It was correctly assumed earlier (from work on grasshoppers by Walter Sutton) that these two chromosomes would be the ones that determined the sex of the organism. Morgan found a fruitfly mutant which had white eyes rather than the usual red ones. When this mutant, a male, was mated with a red-eyed female, all the progeny had red eyes, indicating that the red eyes were dominant. Upon self-breeding of the progeny, the red-eyed flies outnumbered the white-eyed ones by a factor of about three, as would be expected. However, it was found that all white-eyed flies were male. Thus, according to Morgan, the gene for eye color must be on the X chromosome. He was right. This was the first time that anyone had placed a specific gene on a specific chromosome.

Of course, Morgan could not do all this work by himself. He had a large group of graduate students (nicknamed the "fly squad"), some of whom later became independent researchers. One of them was Hermann Muller. He was wondering what caused the mutations. He tried outside changes: no luck: if you cut off the tail of a mouse, all progeny had tails. He tried accidents of all kinds, but no mutations were produced. He then began thinking in terms of the "world of the little". Perhaps mutations came from an ultramiscroscopic accident. Maybe the gene could not escape a speeding electron. He then tried radiation (small, energy-rich particles); he reasoned that there is some natural radiation in the atmosphere, causing rare mutations, and that he might be able to get many more mutations by irradiating his flies with X-rays (invented a couple of years before). He put hundreds of flies in gelatin capsules, irradiated them, and then bred them with untreated flies. The progeny was an interesting mix: there were flies with bulging eyes, flat eyes, dented eyes, purple eyes, yellow eyes, broad wings, curly wings, bumpy wings, etc.. Some were hyper-active, others dopy. Some were big, others small. The conclusion: mutations are caused by ultramiscroscopic collisions, making changes in the hereditary information.

The question now could focus on the chemical identity of the carrier of the hereditary information. As indicated earlier, Miescher had isolated nuclein in large amounts from salmon sperm, and unlike protein, nuclein was found to have a high concentration of phosphorus.

Therefore, it was unlikely that nuclein was just a type of protein. Eventually, chemical analysis revealed that chromosomes are made of both protein and nuclein, and that nuclein is deoxyribonucleic acid (DNA). With both protein and DNA present in chromosomes, which of these carries hereditary information? Almost everyone thought that proteins must be the hereditary material because chemically it is much more complicated than DNA, and the molecule of heredity had to be able to contain an extraordinary amount of information. DNA seemed too simple..... How wrong this reasoning would turn out to be.

In 1928 Frederick Griffith, a scientist who was interested in pathology rather than in genes, made a serendipitous discovery that led to the identification of the chemical composition of genetic material. He was trying to understand the differences between the strain of Diplococcus pneumoniae that cause pneumonia (virulent strains) and the strains that are harmless (non-virulent). The virulent forms had polysaccharide capsules around them that gave them a smooth appearance.

The non-virulent bacteria had no capsules and were rough. Griffith hoped that either heat-killed virulent strains or live non-virulent strains could be used as vaccine. He mixed heat-killed virulent and live non-virulent strains together in hopes of making an effective vaccine. Instead, upon injecting this mixture into mice, the mice died. How could dead bacteria be virulent? He retrieved virulent bacteria from the blood of the dead mice, and he saw that the bacteria were smooth. Something had been transferred from the dead virulent bacteria to the live non-virulent ones. Griffith called this substance the "transforming factor". However, it is unfortunate that Griffith is not given any credit at the MIT web site, and only the people who followed up on it (Avery et al.) are mentioned there.

Griffith did not follow up on identifying the chemical nature of his "transforming factor". This work laid dormant for a while until 1943, when O.T. Avery and colleagues purified the "transforming factor" and found that it was DNA. This finding was viewed with a lot of suspicion by other scientists, as everyone thought that protein (consisting of a long string of subunits, each of which could have 20 different chemical structures) must be the location of hereditary information. DNA (consisting of a much longer string of subunits, each of which could have 4 different chemical structures) was viewed as inadequate to store much information. This way of thinking is typically B.C. (before computers) because we know now that an enormous amount of information can be stored even using a binary (on/off or 0/1) mode.

Developments were rapid from then on. In 1952 the DNA-vs-protein debate was settled conclusively by showing using radioactive isotopes that protein is not passed on from a virus to its progeny, but DNA is. The polynucleotide (DNA) molecule has a number of relatively unusual characteristics for a macromolecule, including the repeated negative charge of the phosphate groups, and the innate affinity of pyrimidine and purine bases to each other. Erwin Chargaff discovered in the forties that the base compositions of DNA from different organisms vary over a relatively wide range (for example, adenine makes up between 23 and 32% of the bases), but that the amount of A consistently was very close to the amount of T, and the amount of C essentially identical to that of G. These two equalities were the first indication that stoichiometric complexation occurs between A and T and between G and C.

Chargaff shared his findings with James Watson, who with the help of molecular models discovered that H-bonded base-paired structures could be formed between A and T, and between G and C, and that these structures have the same overall dimensions. Two H-bonds can be formed between A and T, and three between G and C. Watson brought this information to the attention of Francis Crick, a crystallographer.

Methods had been developed to determine the structure of crystals (repetitively arranged groups of molecules) to atomic resolution, and Crick was an expert in using this technique. This culminated in the elucidation of the DNA structure in 1953.

The X-ray diffraction pattern of crystallized DNA could be interpreted in terms of a helix (essentially a winding staircase) composed of two polynucleotide strands with H-bonded pairs formed between the bases of opposing strands.

Molecular-modeling studies showed that the H-bonded base pairs could form only when the direct-ional senses of the two interacting chains were opposite (anti-parallel). Also, the DNA duplex was found to be "folded" in a three-dimensional structure, a double-helix (like two helices folded into each other). Watson and Crick Crick (but not Chargaff... life not always is fair) got the Nobel Prize in 1962. The Nobel Committee noted that the discovery had "no immediate practical application, but determining the molecular structure of the substance that is responsible for the forms that life takes is a discovery of tremendous importance". The understatement of the year? Without some of these breakthroughs, molecular biology and biotechnology might still be in their infancy.

Antibiotics and Proteins Made by Bacteria

The rapidly increasing resistance of selected microbes against a large number of antibiotics has become a very serious issue worldwide.

Antimicrobial Resistance

Since their discovery during the 20th century, antimicrobial agents (antibiotics and related medicinal drugs) have substantially reduced the threat posed by infectious diseases. The use of these "wonder drugs", combined with improvements in sanitation, housing, and nutrition, and the advent of widespread immunization programmes, has led to a dramatic drop in deaths from diseases that were previously widespread, untreatable, and frequently fatal. Over the years, antimicrobials have saved the lives and eased the suffering of millions of people. By helping to bring many serious infectious diseases under control, these drugs have also contributed to the major gains in life expectancy experienced during the latter part of the last century. These gains are now seriously jeopardized by another recent development: the emergence and spread of microbes that are resistant to cheap and effective first-choice, or "first-line" drugs. The bacterial infections which contribute most to human disease are also those in which emerging and microbial resistance is most evident: diarrhoeal diseases, respiratory tract infections, meningitis, sexually transmitted infections, and hospital-acquired infections. Some important examples include penicillin-resistant Streptococcus pneumoniae, vancomycin-resistant enterococci, methicillin-resistant Staphylococcus aureus, multi-resistant salmonellae, and multi-resistant Mycobacterium tuberculosis. The development of resistance to drugs commonly used to treat malaria is of particular concern, as is the emerging resistance to anti-HIV drugs.

Consequences

The consequences are severe. Infections caused by resistant microbes fail to respond to treatment, resulting in prolonged illness and greater risk of death. Treatment failures also lead to longer periods of infectivity, which increase the numbers of infected people moving in the community and thus expose the general population to the risk of contracting a resistant strain of infection. When infections become resistant to first-line antimicrobials, treatment has to be switched to second-or third-line drugs, which are nearly always much more expensive and sometimes more toxic as well, e.g. the drugs needed to treat multidrug-resistant forms of tuberculosis are over 100 times more expensive than the first-line drugs used to treat non-resistant forms.

In many countries, the high cost of such replacement drugs is prohibitive, with the result that some diseases can no longer be treated in areas where resistance to first-line drugs is widespread. Most alarming of all are diseases where resistance is developing for virtually all currently available drugs, thus raising the spectre of a post-antibiotic era. Even if the pharma-ceutical industry were to step up efforts to develop new replacement drugs immediately, current trends suggest that some diseases will have no effective therapies within the next ten years.

Causes

Microbes (the collective term for bacteria, fungi, parasites, and viruses) cause infectious diseases, and antimicrobial agents, such as penicillin, streptomycin, and more than 150 others, have been developed to combat the spread and severity of many of these diseases. Resistance to antimicrobials is a natural biological phenomenon that can be amplified or accelerated by a variety of factors, including human practices.

The use of an antimicrobial for any infection, real or feared, in any dose and over any time period, forces microbes to either adapt or die in a phenomenon known as "selective pressure". The microbes which adapt and survive carry genes for resistance, which can be passed on.

Bacteria are particularly efficient at enhancing the effects of resistance, not only because of their ability to multiply very rapidly but also because they can transfer their resistance genes, which are passed on when the bacteria replicate. In the medical setting, such resistant microbes will not be killed by an antimicrobial agent during a standard course of treatment. Resistant bacteria can also pass on their resistance genes to other related bacteria through "conjugation", whereby plasmids carrying the genes jump from one organism to another. Resistance to a single drug can thus spread rapidly through a bacterial population.

When anti-microbials are used incorrectly-for too short a time, at too low a dose, at inadequate potency; or for the wrong disease-the likelihood that bacteria and other microbes will adapt and replicate rather than be killed is greatly enhanced. Much evidence supports the view that the total consumption of antimicrobials is the critical factor in selecting resistance. Paradoxically, underuse through lack of access, inadequate dosing, poor adherence, and substandard anti-microbials may play as important a role as overuse. For these reasons, improving use is a priority if the emergence and spread of resistance are to be controlled.

Unprecedented Trends

In the past, medicine and science were able to stay ahead of this natural phenomenon through the discovery of potent new classes of antimicrobials, a process that flourished from 1930-1970 and has since slowed to a virtual standstill, partly because of misplaced confidence that infectious diseases had been conquered, at least in the industrialized world.

In just the past few decades, the development of resistant microbes has been greatly accelerated by several concurrent trends. These have worked to increase the number of infections and thus expand both the need for antimicrobials and the opportunities for their misuse. Such trends include:

- urbanization with its associated overcrowding and poor sanitation, which greatly facilitate the spread of such diseases as typhoid, tuberculosis, respiratory infections, and pneumonia;
- pollution, environmental degradation, and changing weather patterns, which can affect the incidence and distribution of infectious diseases, especially those, such as malaria, that are spread by insects and other vectors;
- demographic changes, which have resulted in a growing proportion of elderly people needing hospital-based interventions and thus at risk of exposure to highly resistant pathogens found in hospital settings;
- the AIDS epidemic, which has greatly enlarged the population of immunocompromised patients at risk of numerous infections, many of which were previously rare;
- the resurgence of old foes, such as malaria and tuberculosis, which are now responsible for many millions of infections each year;

- the enormous growth of global trade and travel which have increased the speed and facility with which both infectious diseases and resistant microorganisms can spread between continents.

As the number of infections and the corresponding use of antimicrobials have increased, so has the prevalence of resistance. In addition, the enhanced food requirements of an expanding world population have led to the widespread routine use of antimicrobials as growth promoters or preventive agents in food-producing animals and poultry flocks. Such practices have likewise contributed to the rise in resistant microbes, which can be transmitted from animals to man.

Factors that Encourage the Spread of Resistance

The emergence and spread of antimicrobial resistance are complex problems driven by numerous interconnected factors, many of which are linked to the misuse of antimicrobials and thus amenable to change. In turn, antimicrobial use is influenced by an interplay of the knowledge, expectations, and interactions of prescribers and patients, economic incentives, characteristics of a country's health system, and the regulatory environment.

Patient-related factors are major drivers of inappropriate antimicrobial use. For example, many patients believe that new and expensive medications are more efficacious than older agents. In addition to causing unnecessary health care expenditure, this perception encourages the selection of resistance to these newer agents as well as to older agents in their class. Self-medication with antimicrobials is another major factor contributing to resistance. Self-medicated antimicrobials may be unnecessary, are often inadequately dosed, or may not contain adequate amounts of active drug, especially if they are counterfeit drugs. In many developing countries, antimicrobials are purchased in single doses and taken only until the patient feels better, which may occur before the pathogen has been eliminated. Inappropriate demand can also be stimulated by marketing practices. Direct-to-consumer advertising allows pharmaceutical manufacturers to market medicines directly to the public via television, radio, print media, and the Internet. In particular, advertising on the Internet is gaining market penetration, yet it is difficult to control with legislation due to poor enforceability. Prescribers' perceptions regarding patient expectations and demands substantially influence prescribing practice. Physicians can be pressured by patient expectations to prescribe antimicrobials even in the absence of appropriate indications. In some cultural settings, antimicrobials given by injection are considered more efficacious than oral formulations. Such perceptions tend to be associated with the over-prescribing of broad-spectrum injectable agents when a narrow-spectrum oral agent would be more appropriate. Prescribing "just to be on the safe side" increases when there is diagnostic uncertainty, lack of prescriber knowledge regarding optimal diagnostic approaches, lack of opportunity for patient follow-up, or fear of possible litigation. In many countries, antimicrobials can be easily obtained in pharmacies and markets without a prescription. Patient compliance with recommended treatment is another major problem. Patients forget to take medication, interrupt their treatment when they begin to feel better, or may be unable to afford a full course, thereby creating an ideal environment for microbes to adapt rather than be killed. In some countries, low quality antibiotics (poorly formulated or manufactured, counterfeited or expired) are still sold and used for self-medication or prophylaxis.

Hospitals are a critical component of the antimicrobial resistance problem worldwide. The combination of highly susceptible patients, intensive and prolonged antimicrobial use, and cross-infection has resulted in nosocomial infections with highly resistant bacterial pathogens. Resistant hospital-acquired infections are expensive to control and extremely difficult to eradicate. Failure to implement simple infection control practices, such as handwashing and changing gloves before and after contact with patients, is a common cause of infection spread in hospitals throughout the world. Hospitals are also the even-tual site of treatment for many patients with severe infections due to resistant pathogens acquired in the community. In the wake of the AIDS epidemic, the prevalence of such infections can be expected to increase.

Veterinary prescription of antimicrobials also contributes to the problem of resistance. In North America and Europe, an estimated 50% in tonnage of all antimicrobial production is used in food-producing animals and poultry. The largest quantities are used as regular supplements for prophylaxis or growth promotion, thus exposing a large number of animals, irrespective of their health status, to frequ-ently subtherapeutic concentrations of antimicrobials. Such widespread use of antimicrobials for disease control and growth promotion in animals has been paralleled by an increase in resistance in those bacteria (such as Salmonella and Campylobacter) that can spread from animals, often through food, to cause infections in humans.

The Need for a Global Response

In September 2001, WHO launched the first global strategy for combating the serious problems caused by the emergence and spread of antimicrobial resistance. Known as the WHO Global Strategy for Containment of Antimicrobial Resistance, the strategy recognizes that antimicrobial resistance is a global problem that must be addressed in all countries. No single nation, however effective it is at containing resistance within its borders, can protect itself from the importation of resistant pathogens through travel and trade. Poor prescribing practices in any country now threaten to undermine the potency of vital antimicrobials everywhere.

The strategy recommends interventions that can be used to slow the emergence and reduce the spread of resistance in a diverse range of settings. The interventions are organized according to groups of people whose practices and behaviours contribute to resistance and where changes are judged likely to have a significant impact at both national and international levels. These include consumers, prescribers and dispensers, veterinarians, and managers of hospitals and diagnostic laboratories as well as national governments, the pharmaceutical industry, professional societies, and international agencies. Global principles for the containment of antimicrobial resistance in food-producing animals were issued by WHO in June 2000. As much of the responsibility for containing resistance rests with national governments, the strategy gives particular attention to interventions involving the introduction of legislation and policies governing the development, licensing, distribution, and sale of antimicrobial agents. The strategy is sufficiently flexible to be applied in poor and wealthy nations alike. The process for selecting the necessary interventions to limit emerging antimicrobial resistance can be based on the diseases most prevalent in a given country. In advocating widespread adoption of this strategy, WHO aims to encourage the urgent actions needed to reverse or at least curtail trends which have major economic as well as health implications. Moreover, in view of the global nature of the antimicrobial resistance problem, the efforts of any nation to implement the WHO Global Strategy are likely to be felt worldwide.

The strategy builds on a number of WHO activities aimed at both monitoring the global emergence and spread of antimicrobial resistance and extending direct support to countries. WHO helps countries establish laboratory-based networks for the surveillance of resistance. Specific activities include staff training, support in methods for the quality assurance of laboratory tests, and provision of laboratory reagents. In addition, WHO distributes a computer software program, WHONET. Microbiologists, clinicians, and infection control workers may use this software to improve the systematic monitoring of drug resistance in their hospitals and communities and to share their data in a common format among national networks.

Since 1977, WHO has produced Model Lists of Essential Drugs in order to help governments select the most effective and appropriate drugs in line with priority needs. The lists, which are regularly revised, also contribute to the rational purchasing and use of drugs. Studies have demonstrated that in those areas in which an essential drugs programme is in operation, significantly more essential drugs are available, significantly fewer injections and antimicrobials are utilized, and drug stocks last about three times longer than in regions without such a programme. At present over 120 countries have implemented an essential drugs list.

With the first global strategy for containment of antimicrobial resistance now available, WHO is also in a position to advise health policy-makers and managers on the specific interventions needed to safeguard the effectiveness of vital drugs and thus ensure that their life-saving capacity remains available to future generations.

Questions about Antibiotic Resistance

What are bacteria and viruses?

Bacteria are single-celled organisms usually found all over the inside and outside of our bodies, except in the blood and spinal fluid. Many bacteria are not harmful.

In fact, some are actually beneficial. However, disease-causing bacteria trigger illnesses, such as strep throat and some ear infections. Viruses are even smaller than bacteria. A virus cannot survive outside the body's cells. It causes illnesses by invading healthy cells and reproducing.

What kinds of infections are caused by viruses and should not be treated with antibiotics?

1. Colds
2. Flu
3. Most coughs and bronchitis
4. Sore throats (except for those resulting from strep throat)

How do I know when an illness is caused by a viral or bacterial infection?

Sometimes it is very hard to tell. Consult with your doctor to be sure.

When do I need to take antibiotics?

Antibiotics are very powerful medications. They should only be used when prescribed by a doctor to treat bacterial infections. Do I need an antibiotic when mucus from the nose changes to yellow or green?

Yellow or green mucus does not indicate a bacterial infection. It is normal for the mucus to get thick and change color during a viral cold.

Should I ask my doctor to prescribe antibiotics?

Talk to your doctor about the best treatment. You should not expect to get a prescription for antibiotics. If you have a viral infection, antibiotics will not cure it, help you feel better, or prevent someone else from getting your virus.

What is antibiotic resistance and why should I be concerned?

Antibiotic resistance occurs when bacteria change in a way that reduces or eliminates the effectiveness of antibiotics. These resistant bacteria survive and multiply-causing more harm, such as a longer illness, more doctor visits, and a need for more expensive and toxic antibiotics. Resistant bacteria may even cause death.

What can I do to avoid antibiotic-resistant infections?

Start by talking with your healthcare provider about antibiotic resistance.

Ask whether an antibiotic is likely to be effective in treating your illness.

Do not demand an antibiotic when your healthcare provider determines one is not appropriate.

Ask what else you can do to help relieve your symptoms.

What can I do to protect my child from antibiotic-resistant bacteria?

Use antibiotics only when your doctor has determined that they are likely to be effective. Antibiotics will not cure most colds, coughs, sore throats, or runny noses. Children fight off colds on their own.

If mucus from the nose changes from clear to yellow or green, does this mean that my child needs an antibiotic?

Yellow or green mucus does not mean that your child has a bacterial infection. It is normal for the mucus to get thick and change color during a viral cold.

Does this mean that I should never give my child antibiotics?

Antibiotics are very powerful medicines and should only be used to treat bacterial infections. If an antibiotic is prescribed, make sure you take the entire course and never save the medication for later use.

How do I know if my child has a viral or bacterial infection?

Ask your doctor. If you think that your child might need treatment, you should contact your doctor. But remember, colds are caused by viruses and should not be treated with antibiotics.

Facts about Antibiotic Resistance

Antibiotic resistance has been called one of the world's most pressing public health problems. The number of bacteria resistant to antibiotics has increased in the last decade. Nearly all significant bacterial infections in the world are becoming resistant to the most commonly prescribed antibiotic treatments.

Every time a person takes antibiotics, sensitive bacteria are killed, but resistant germs may be left to grow and multiply. Repeated and improper uses of antibiotics are primary causes of the increase in drug-resistant bacteria. Misuse of antibiotics jeopardizes the usefulness of essential drugs. Decreasing inappropriate antibiotic use is the best way to control resistance.

Children are of particular concern because they have the highest rates of antibiotic use. They also have the highest rate of infections caused by antibiotic-resistant pathogens.

Parent pressure makes a difference. For pediatric care, a recent study showed that doctors prescribe antibiotics 65% of the time if they perceive parents expect them; and 12% of the time if they feel parents do not expect them.

Antibiotic resistance can cause significant danger and suffering for people who have common infections that once were easily treatable with antibiotics. When antibiotics fail to work, the consequences are longer-lasting illnesses; more doctor visits or extended hospital stays; and the need for more expensive and toxic medications. Some resistant infections can cause death.

Prevention of Antibiotic Resistance

Do not take an antibiotic for a viral infection like a cold, a cough or the flu.

Take an antibiotic exactly as the doctor tells you. Do not skip doses. Complete the prescribed course of treatment, even if you are feeling better.

Do not save any antibiotics for the next time you get sick. Discard any leftover medication once you have completed your prescribed course of treatment.

Do not take antibiotics prescribed for someone else. The antibiotic may not be appropriate for your illness. Taking the wrong medicine may delay correct treatment and allow bacteria to multiply.

Antibiotic prescriptions in outpatient settings can be reduced dramatically-without adversely affecting patient health-by not prescribing antibiotics for viral illnesses, such as colds, most sore throats, coughs, bronchitis, and the flu.Parents should not demand antibiotics when a healthcare provider has determined they are not needed.

Parents should talk with their healthcare provider about antibiotic resistance.Parents should not give their children antibiotics for a viral infection like a cold, a cough, or the flu. Antibiotics should be used only to treat bacterial infections. Parents should ensure that their children take all medication as prescribed, even if symptoms disappear. If treatment stops too soon, some bacteria may survive and re-infect.

A Prescription for Parents: Five Hints to Understanding Antibiotic Usage. When are antibiotics necessary? Your doctor can best answer this complicated question and the answer depends on the diagnosis. Here are a few examples:

1. *Ear Infections:* There are several types; many need antibiotics, but some do not.
2. *Sinus Infections:* Most children with thick or green mucus do not have sinus infections. Antibiotics are needed for some long-lasting or severe cases.
3. *Cough or Bronchitis:* Children rarely need antibiotics for bronchitis.
4. *Sore Throat:* Viruses cause most cases. Only one major kind, "strep throat," requires antibiotics. This condition must be diagnosed by a laboratory test.

5. *Colds:* Colds are caused by viruses and may last for two weeks or longer. Antibiotics have no effect on colds, but your doctor may have suggestions for obtaining comfort while the illness runs its course.

It is worth noting that viral infections sometimes lead to bacterial infections. But treating viral infections with antibiotics will not prevent bacterial infections and may trigger infections with resistant bacteria. Keep your doctor informed if the illness gets worse, or lasts a long time, so that the proper treatment can be given as needed.

Fluid in the Middle Ear

A doctor said your child has fluid in the middle ear, also called otitis (oh-TIE-Tus) media with effusion (uh-FEW-zhun) (OME). Fluid usually does not bother children and it almost always goes away on its own. This does not have to be treated with antibiotics, unless it lasts for a few months. Here are some facts about OME and ear infections.

What are the main kinds of ear infections?

Swimmer's ear (otitis externa) is an infection of the ear canal that can be painful and is treated with eardrops.

A middle ear infection, which a doctor might call "acute otitis media" (AOM), may cause ear pain, fever, or an inflamed eardrum, and is often treated with oral antibiotics.

What causes OME?

Fluid may build up in the middle ear for two reasons. When a child has a cold, the middle ear makes fluid just as the nose does- it just doesn't run out as easily from the middle ear. After a middle ear infection, fluid may take a month or longer to go away.

Are antibiotics ever needed for OME?

Sometimes antibiotics may be needed if the fluid is still present after a few months and is causing decreased hearing in both ears. For this reason, your child will need an ear check in a few months. If there is still fluid in the middle ear, your child may need a hearing test.

What should I do?

The best treatment is to wait and watch your child. Since fluid in the middle ear rarely bothers children, it is best to let it go away on its own. Right now, your child might not need antibiotics. You may need to schedule a visit to see the doctor again in a few months to be sure the fluid is gone.

Why not try antibiotics now?

Taking antibiotics when they are not needed can be harmful. Each time people take antibiotics, they are more likely to carry resistant germs in their noses and throats. Common antibiotics cannot kill these resistant germs. Your child may need antibiotics that are more costly, given by a needle, and/or administered in the hospital. Since OME will almost always get better on its own, it is better to wait and take antibiotics only when they are needed.

A Child's Runny Nose

Your child has a runny nose. This is a normal part of what happens during the common cold and as it gets better. Here are some facts about colds and runny noses.

What causes a runny nose during a cold?

When germs that cause colds first infect the nose and sinuses, the nose makes clear mucus. This helps wash the germs from the nose and sinuses. After two or three days, the body's immune cells fight back, changing the mucus to a white or yellow color. As the bacteria that live in the nose grow back, they may also be found in the mucus, which changes the mucus to a greenish color. This is normal and does not mean your child needs antibiotics.

What should I do?

The best treatment is to wait and watch your child. Runny nose, cough, and symptoms like fever, headache, and muscle aches may be bothersome, but antibiotics will not make them go away any faster. Some people find that using a cool mist vaporizer or saltwater nose drops makes their child feel better.

Are antibiotics ever needed for a runny nose?

Antibiotics are needed only if your doctor tells you that your child has sinusitis. Your child's doctor may prescribe other medicine or give you tips to help with a cold's other symptoms like fever and cough, but antibiotics are not needed to treat the runny nose.

Why not try antibiotics now?

Taking antibiotics when they are not needed can be harmful. Each time people take antibiotics, they are more likely to carry resistant germs. Your child may need antibiotics that are more costly, given by a needle, and/or administered in the hospital. Since a runny nose almost always gets better on its own, it is better to wait and take antibiotics only when they are needed.

Cold and Flu Season

Colds, flu, and most sore throats and bronchitis are caused by viruses. Antibiotics do not help fight viruses. And they may do more harm than good: taking antibiotics when they are not needed-and cannot treat the illness-increases the risk of a resistant infection later.

Antibiotics Are Not for Colds and Flu.

Most infections are caused by two main types of germs-bacteria and viruses. Bacteria are organisms found almost anywhere, except normally sterile sites, such as the blood stream and spinal fluid. A few bacteria, known as pathogens, can cause diseases in humans, animals, and plants.

Viruses are organisms that cause disease by invading healthy host cells. As virus particles multiply, the host cells burst, allowing the viruses to infect other cells.

Cough and Cold Medicines for Children

What can parents do if their children are too young or the healthcare provider advises against using cough and cold medicines? Parents might consider clearing nasal congestion in infants with a rubber suction bulb. Also, secretions can be softened with saline nose drops or a cool-mist humidifier.

Are cough and cold medicines safe for children under 2 years of age? There are no Food and Drug Administration (FDA)-approved dosing recommendations for children under 2 years of age. These drugs can, in rare cases, be harmful or even fatal. Parents and healthcare providers should use caution when giving cough and cold medicines to children under 2 years of age.

Do cough and cold medicines work in children under 2 years of age? There is little evidence that cough and cold medicines work in children under 2 years of age.

Should parents give cough and cold medicines to children under 2 years of age?

Parents should consult a healthcare provider before giving cough and cold medicines to their children and should always tell providers about all prescription and over-the-counter medicines they are giving their child.

Should healthcare providers prescribe cough and cold medicines to children under 2 years of age?

Healthcare providers should exercise caution when recommending or prescribing cough and cold medicines to children under 2 years of age and should always ask caregivers about any other cough and cold medicines the child might be receiving. No FDA-approved dosing recommendations exist for over-the-counter cough and cold medicines in children under 2 years of age.

What should parents and doctors be careful of if they want to give cough and cold medicines to children under 2 years of age? Be especially careful if giving more than one cough and cold medicine at a time to children under 2 years of age. Two medicines may have different brand names but may contain the same ingredient. Some cough and cold medicines contain more than one active ingredient.

9

Biotechnology in Medicinal Agricultural Development

Introduction

We are at the threshold of 21st century, where great challenges on three fronts namely, production, pollution and population are awaiting the developing countries. Moreover, in the era of privatization, globalization and liberalization, there is heavy international competition within as well as between the countries on all fronts including the agriculture. The present scenario indicates that there would be "survival of the fittest". Needless to mention that by 2000 AD our population will be around one billion the priority is clear-cut before us. We have to climb upward to achieve higher and higher production to ensure that every mouth of our ever increasing population has enough to eat.

The quest for agriculturists for greater productivity and improvement of existing cultivars with better quality food continues with great vigour and spirit. The conventional methods of plant breeding and traditional agricultural practices have done a tremendous job and contributed to a great deal towards the above goal.

However, in view of the acuteness of the problem and renewed fears regarding the availability of the proper and enough food, these methods alone are not sufficient to meet the situation. Although, the FAO believes that food production will continue to increase for the next 20 to 25 years at the same rate as experienced during the past 30 years to meet the futuristic demand.

The worldwatch Institute has, however, reported that this rate may not be sustainable. Moreover, they believe that science and technology can no longer ensure the onward march of achieving higher and higher yields. In addition we are also subjected to an almost daily litany of doom and despair as a result of global warming, depletion of ozone layer, oil and water running out and so on. The next century certainly packed with trouble if the meteor does not get us first, then we will gradually either be roasted, frozen or desiccated, and to make doubly sure, shall certainly be starved as well.

It seems that if we have to stand any chance whatsoever in the next century, our capacity to adopt the living resources will be extremely important for our survival. Because of changes in the environment and depleting energy resources, this capacity to adapt will be every bit as important to those areas of the world which are blessed with overproduction like Europe today. Accelerated adaptation through genetic change is, of course, one of the important means by which plant breeders have achieved their aims in the past. Today the new techniques of biotechnology in general and genetic engineering in particular may offer to the plant breeders and agriculturists the chance of speeding up adoption to an extent hitherto considered impossible.

This new power will undoubtedly bring risks but it could also be the only chance we have for escaping the dangers ahead. Biotechnology being a multidisciplinary subject requires coordinated efforts and expertise of scientists with different backgrounds to put together in order to achieve success. It may be defined as may technique that uses living organisms, or parts of organisms to make or modify products, to improve plants or animals or to develop microorganism for specific uses. It includes well established technologies such as those used in breeding, plant propagation and conventional animal vaccine productions. Modern biotechnology encompasses the more recently developed techniques involving the use of recombinant DNA technology, monoclonal antibodies and new techniques of cell culture. Recent development in biotechnology have made it possible to move genes from microbes, plants and animal species into plants of interest. Biotechnology helps in the commercial utilization of gene and other things for the benefit of human kind. However, destruction of ecosystems has been a major concern of biotechnology.Developing countries are naturally attracted to the potential applications of biotechnological research in solving problems of hunger, energy supply and improving the quality of life. The priorities of different countries however very widely, In this context, the National Biotechnology Board of India has chosen genetic engineering, photosynthesis, tissue culture, enzyme engineering, alcohol fermentation and immuno-technology as areas of immediate interest.

Application of Cell Biotechnology

Cell biotechnology or plant tissue culture is an enabling technology from which many novel tools have been developed to assist plant breeders. These tools can be used to increased the efficiency of the breeding process, to improve the accessibility of existing germplasm and to create new variation for crop improvement. They include micropropagation, another culture, in vitro selection, embryo rescue, somaclonal variation, somatic bybridization and transformation. Of these somaclonal variation occupies a somewhat unique position, because it has both the advantages and disadvantages of tissue culture systems.

Somaclonal variation was first defined as such by Larkin and Scowcroft who reviewed the subject in 1981 and were among several authors at that time to draw attention to its potential use for crop improvement (Larkin and Scowcroft, 1981). An extensive number of reports soon followed in a whole range of species, indicating that somaclonal variation was widespread, and therefore, ostensibly accessible to all plant breeders (Karp, 1991).

The Utilization of new genetic variability has become one of the major objectives of tissue culture, The assembly of genetic variability is vital for improvement of crop plants. Somaclones for the resistance of downy mildew. Fiji disease and eye spot disease were identifies. Application of somaclonal variation in crop improvement can produce increased yield and resistance to biotic as well as abiotic a stresses. Some of the crops where somaclones were produced are rice, wheat, maize, tomato, geranium, sweet pototo, sugarcane, rice, celery and brown mustard (Table 1). Our country research efforts have developed a variety called 'Pusa Jaikisan' in brown mustard evolved by somaclonal variation.

Micropropagation is another technique of great significance to agriculture which is based on in vitro culture of plant cell. Each and every cell of a plant has the potency to be regenerated into complete plant by tissue culture techniques and this property of the cell is known as totipotency. Using this technique hundreds and thousands of seedlings can be generated in a very short time, starting from a limited number of explants. This methods of multiplication has two major advantages. First is the multiplication of new elite cultivars. For example, after its release, a new sugarcane cultivar may take 5-10 years before sufficient planting material may be generated for wide coverage in the farmers field. But using micropropagation techniques a comparable level of multicipation may be achieved within 2 years. The second advantage is that we get disease free planting material because of the use of aseptic conditions during tissue culture. There is also reports of rejuvenation and increased vigous following tissue culture. This can also help in rescue of important materials which are on the verge of extinction due to disease attack. For example, a rare variety of banana used for temple offering in Karnataka has been rescued through tissue culture techniques.

Micropropogation of fruit and ornamental plants provides a useful avenue for the employment generation for rural youths at cottage level industries. In species like Daffodies and gladioli, microporpagation techniques are being used to speed up the release of new varieties. In strawberry, millions of plants can be produced from a single mother plant in one year.

Micropropagation work undertaken at our Defence Agricultural Research Laboratory. Pithograh was resulted into successful multiplication of strawberry, petunia, carnation and tomato hybrid. In case of strawberry a well adapted variety of hill vi., Jyolikit Selection was taken for micropropagation studies with objective to large scale production of disease free propagules since this fruit plant is highly susceptible to number of fungal, bacterial and viral diseases. The result showed that internodal segments are most suitable explant material for micropropagation under in Vitro condition. MS formulated with 1ppm of BAP and 0.1ppm of GA3 was found optimal protocal for maximizing shoot proliferation in explant where MS media modified with 1 ppm 1AA was most suitable for early rooting in shootlets. Peat moss was found to be most suitable potting culture for hardening of tissue culture raised plants of strawberry var. Jyolikot Selection.

Micropropagation studies in Petunia Hybrida were undertaken in order to raise the plant seedlings in mass. The result showed that optimum number of shootlets could be obtained after culturing the leaf explant of size 5mm X 2.5mm in MS media supplemented with 1ppm of BAP + 0.5 ppm of IAA with an inclubation period of 34 days. Protocol standardisation revealed that MS medium supplemented with 1 ppm of IBA was suitable for an early and healthy rooting in tissue culture raised shootlets. Hardening process was most successful with high relative humidity (90-95%), Temperature 23 + 10 C and potting mixture of sand and soil in equal proportion. Work on micropropagation of carnation was resulted into stan-dardization of an vitro culture media for shoot and root proliferation. Protocol found most promising for shoot proliferation was MS medium supplemented with 1 ppm of BAp AND 0.1 ppm of IAA where MS med-ium supplemented with 1 ppm of NAA was identified to most suitable for rooting in shootlets.

Micropropagation studies were also undertaken in hybrid tomato to generate large scale seedings to fulfil the need of army personnel as well as local growers with pure and healthy planting material. Based on the observation it was found that tissue culture medium comprised of MS,0.5 ppm kinetin, 0.5 ppm IAA and 0.3 ppm GA3 was most suitable for shoot growth and MS medium supplemented with ppm of IAA or MS with 1 ppm of IBA was most suitable for early and healthy rooting in the explant material. Tissue culture regenerated plantlets showed high survival under relative humidity varied from 85-95 per cent, temperature 25+20C and potting mixture sand and oak forest humus in equal ratio.

Application of DNA Technology

Genetic engineering opens a totally new dimension for bioprospecting. The search for new genes and their application is the primary objective of the biotech industry. Gene technology now enable humans to integrate revolutionary new properties in to cultivated plants through inter-specific or inter-generic gene transfer which was not possible through classical approach of crop improvement.

Genetic engineering is the direct introduction into a plant of an isolated or modified single gene using transformation techniques. Now it is possible to transform almost all of the plants cultivated by man. Where this has not happened, it is more to do with the insignificance of the plant in agriculture or forestry rather than a reflection of a stubborn resis-tance to all attempts to transform it. A couple of years ago, cereals were regarded as being recalcitrant species. This has new changed and transformation has been reported as being successful for nearly all cereal species.

Transgenic Plants

Transgenic is used to describe plants which have had DNA intro-duced into them by means other than by the transfer of DNA from a sperm cell to an egg. Development achieved through modern biotechnology in the form of transgenics have given new options of controlling insects, diseases and weed pests, while improving the overall integrity and consequences of agricultural practices. Biotechnology combined with modern crop management techniques and the responsible use of pesticides gives the best tools available to ensure healthy, high yielding crops.

However, the benefit expected from the release of genetically modified organisms into environment are quite more which are seems to pay major role in bioremediation, environmental improvement, agriculture, food industry and health care. A number of transgenic plants in various crop species covering wide range of altered characteristics have been approved for commercial cultivation.

Resistant to Biotic Stresses

Damage to crops by biotic factors like insects-pests and weeds is a major limiting factor in agriculture economic in tropical and temperate regions of the world. Despite large scale investment in the chemical control of pests, they cause great economic loss by damaging or destroying the crops. With a tremendous development in techniques to engineer crops genetically, we now have the ability to make broader use of natural insecticides.

The choice of a Bt endotoxin, as the first insecticidal protein for introduction into plants, was based on the extensive knowledge gathered about this class of crystal protein since 1902 and such a strategy has been successfully achieved.

The application of glyphosate kills crop plants just as effectively as it kills weeds. The usefulness of glyphosate as a weed control agent in agric-ulture could be enhanced if resistance to herbicide could be selectively engineered into crop plants. Transgenics have been developed by introducing the corresponding resistance genes from weeds and microbes into crop plants such as tobacco.

In addition genes conferring resistance against herbicides, viz., Sulphobylurea, Imidazolinones, Phosphinothricin, Asulam, Bromoxynil and 2,4-D have been engineered and transgenic plants developed.

Abiotic Resistance

A large area of our country is under stress conditions i.e. saling, alkaline water, cold and heat stress etc. transgenics produced from introducing mtl 1D gene from E. coli in tobacco and arabidopsis showed tolerance to high leave of NaCl. Thermoc tolerance was found in Arabidopsis plants engineered with hsf gene. It has been reported that over-expression of sac B gene from Bacillus subtillis leads to high level of fructans in tobacco cells, and this is associated with increased drought tolerance whereas chemically synthesized antifreeze protein genelala 3 leads to improved freezing tolerance of tobacco plants.

Genetic engineering for improved tolerance to abiotic stresses is, therefore, the need of the hour as the existing cultivars in most cases are capable of giving much higher biomass that what we harvest today. While it is true that the response of plants to these stresses is multigenic, the recent success achieved in genetic engineering against excess salt, high and low temperature as well as less or excess water by altering individual gene is noteworthy.

Product Quality

A group of scientists have reported the ethylene synthesis in transgenic tomato plants, using anti-sense to a gene that has been identified as ACC oxidase. The enzyme convert ACC to ethylene. Ethylene production was inhibited by 97% with a signification reduction in over-ripening and shrivelling. Flavr Savr and Endless Summer in tomato, Freedom II in squash, high lauric acid reposed (Canola) and Round Up Ready soybean are some examples of the crops that already being commercially grown in developed countries.

Genetic engineering of metabolic pathways of fruits and vegetables has therefore given control over post-harvest process which not only leads to improvement of quality characteristics but also enhances processability, transportability and prologation of shelf like.

In addition, engineered plants with change fatty acid composition of edible oil, reduced antinutritional components in many food crops, production of palm oil or animal derived industrial oil into plants like Brassica are some of the good examples of novel products that can be expected from biotechnology. Significant progress has been achieved for prod-uction of biodegradable plastics derived from biomsass of genetically altered plants having gene from bacteria.

With the advancement in DNA technology and understanding the structure and function of gene and its products more and more transgenics are coming out from various research programme all over the world. However, boon of biotechnology has been confined to few developed countries and developing countries are still in vogue to harvest the benefits of modern technology.

More that 48 transgenic has picked up remarkably in last couple of years. During 1986-87, 25000 field trials were conducted with transgenics of more that 60 crops in 45 countries. Sixty per cent of these trials were in first 10 years and 40% in the last two years.

Area under transgenics cultivation has also increased tremendously. During 1996, the transgenic crops covered an area of 2.8 million has which increased about 4 times during 1997 and 10 times during 1998.

Table: *Area of transgenic crops planted (m ha)*

Country	***1997***	***1998***
USA	8.1	20.5
Argentina	1.4	4.3
Canada	1.3	2.8
Australia	0.1	0.1
Mexico	<0.1	0.1
Spain	0.0	<0.1
France	0.0	<0.1
South Africa	0.0	<0.1
Total	11.0	27.8

Source: Anne Simon Maffat (1998)

Conclusion

A major challenge for agriculture in the 21st century in developing nations would be to speed up production process economically to fulfil the aspirations of huge populace, to achieve diversification and adding value to the primary produce so as to make agriculture enterprise farmers as well as environmental friendly. Biotechnology based advance technologies are expected to materialize many of our expectations in the coming millennium.

Viruses

To minimize the losses due to viral disease and to reduce the use of chemicals, the cross protection phenomenon of viral coat protein and movement proteins have been employed. This method has been used to develop transgenic plants which conferred significant levels of resistance in tobacco plants to TMV or to alfa mosaic virus (AMV) using the AMV coat protein gene. In some viruses it has been found that certain sequences termed satellite RNAs act to reduce disease symptoms in infected plants and as such it is being utilized for developing transgenic with virus resis-tance. Antisense RNA which binds to sense RNA and artificial ribozyme molecules which cleave specific gene transcripts were found very powerful tools for the inhibition of viral disease development.

Genetic engineering of antisense or ribozyme genes complementary to viral genes whose products are essential may prove one of the best way of viral disease control.

Insects

Transgenic crop plants that produce pesticidal proteins such as Bt toxin can offer the growers advantage of increased yields, numerous environmental benefits due to decreased use of conventional insecticidals.

A transformed tobacco plants with the gene encoding the cowpea trypsin inhibito, in feeding experiments demonstrated that expression of the gene conferred resistance to a range of insect genera, including Heliothis, Spodoptera, Diabrotica and Tribolium the protein is not however toxic to humans.

Herbicides

The use of herbicidea has become an important feature of modern agriculture as a means of controlling weed species, however, it is economically feasible to produce specific herbicides for use with every individual crop species. A more generic approach would be to transform a range of crop plants with a gene which confers tolerance to a single non-selective herbicide.

Biotechnology for the Cotton Farmer Problems and Prospects

Cotton or white gold as it is aptly called is grown for its lint and seed which yield cotton fibre and seed oil, respectively. This crop occupies 32-33 mha of world area with a production of 20-25 metric tonnes. In India its area spans over 7.5 mha with an average yield of 290 kg/ha of lint and 870 kg/ha of seed cottom.

To meet the challenges of 2000AD with a population of more that 960 million, a total production of 19 million bales is required as against the 13-14 million bales of today. This can be achieved by the use of improved crop production practices coupled with appropriate pest management tactics. In addition, generation of novel. Transgenics may help achieve the near impossible. Genes that have been identified as potentially profitable, if engineered into acceptable cultivars can be used to generate such transgenics. Among these are genes imparting resistance to herbicides, insects, pathogens and abiotic stresses. It is also widely accepted now that a number of other qualitative characters can be improved, such as fibre strength, fineness, colour and thermal adapta-bility of the fibre.Bollworm and herbicide resistant transgenic cotton have received the approval of the Environmental Protection Agency and have been commercially released in the US for cultivation. Consi-dering the fact that numerous biotic and abiotic stresses limit cotton production, it is likely that future strategies might orient towards the development of a multiadversity resistant high yielding transgenic cotton variety with superior fibre qualities.

The most important aspects in the development of transgenic plants are

- Identification, isolation and clining of the desirable genes
- Transformation of plant cells or tissues with a suitable vector
- Regeneration of the transformed cells or tissue
- Confirmation of the gene integration and expression
- Hardening and field establishment of the transgenics.

Major achievements worldover in the field of cotton transgenics

Genes

Genes for Insect Resistance Used

- Toxin genes from Bacillus thuringiensis
- Protease inhibitor genes from plants & the horn worm

Genes used for herbicide resistance :

- Nitrilase gene from a bacteria Klebsiella spp. for Bromoxynil resistance
- Acstolactate synthase gene for resistance against sufonyl urea & imidazoline group
- ERPSP synthase gene for glyphosate resistance.

Genes used to Improve fibre qualities :

- Polyhydroxybutyrate & polyhydroxyacetate from bacteria
- Expression of indigo pigment in fibres

Genes to be Used

- Cholesterol oxidase genes against insects
- Heliothis stunt virus genes against Heliothis

- Genes encoding chitinases glucanases, attacins cecropins vital coat proteins etc. against diseases.

Transformation

- Agrobacterium medicated transformation
- Particle acceleration gene delivery systems have been successfully used all over the world.

Regeneration

- Callus fissue regeneration-mostly genotype specific
- Protoplasts were regenerated successfully.

Meristematic tissues were successfully regenerated.

Expression

- Enhanced expression was achieved using truncated forms of the full length genes, repeat copies of the propoter, modification of the bacterial codons to plant prefered type.

Commercial Release :

- Bollworm resistant transgenics-NUCOTN 33M & NUCOTN 35B were released in the US and cotton transgenics-are in now under field cultivation in Australia, South Africa and China.
- 8 lakh hectares were planted with the transgenic cottons in 1996 and 1997
- Transgenic cotton plants in Australia were also found to be attacked by the bollworms.
- Herbicide (Bromoxynil) resistant transgenics BXN 57 and BXN 53 were released in the US in 1996.

Problems Encountered/Expected

- Gene silencing and expression instability
- Resistance in insects to the toxins used. Studies on baseline toxicity indicated that there is a natural variation in the ability of Helicoverpa armigera in tolerating the CrylAc toxins. Some strains are capable of surviving the totoxins by virtue of an in built capacity to tolerate the toxins, while others are not. While development of Helicoverpa resistance to the CrylAc toxins itself may not be of immediate concern, a slight shift in the tolerance levels resulting from continuous exposure to the transgenics, can cause control difficultes.
- High cost of transgenic seed.

- Microecological changes resulting in pest shifts. CrylAc is a broad spectrum lepidopteran toxins that can cause significant changes in almost all populations of lepidopteran insects occurring on the cotton crop. Some of these insects, whichcause negligible economic damage to the crop, harbour parasitoids which have the potential to keep the bollworm populations under check. Consequently, in the absence or low populaions of natural enemies, the bollworms can emerge as stronger pests that before. Insects pests such as Spodoptera litura which are less affected by the CrylAc toxins and which have been under check due to the use of pyrethroids can resurgace as major pests as pyrethroid use is likely to be reduced on transgenic Bt cottons.

Some Recent Development in Cotton Transgenics

- Three Bt transgenic lines, 95-1, RIOI and R-19 have been indigenously developed in China through Agrobacterium medicated (Somatic embryo) and pollen tube pathway transformation. Thestability of Bt gene expression and field efficacy have been confirmed.
- Microprojectile bombardment parameters for pollen mediated transformation and genetype independent protocols have been develop in China.
- A new methods of transformation through injection of Agrobacterium into developing embryos, has been reported from South Africa.
- Stable transformation and regermation has been reported for Bt transgenics in Uzbekistan, China, Egypt and Australia. Pakistan reports development of cotton transgenic plants resistant torearcurl virus.
- An antisense DNA of CLCUV DNA-A bome ACI gone along with the antesense DNA of the AC2 and AC3 gene was used for the vector construction and transgenic cotton resistant to the CLCUV was developed in Pakistan.
- In Russia, A glucanase gene has been isolated from thermo philic bacteria and is being used as a new reporter system fortransgenic cotton development.
- Genetically engineered Baculovirus ACMNPV with genes from straw itch mite has been developed been developed by Monsanto and Zeneca with researchers from Madison, USA.

- Bt rransgenic cotton currently occupies 20% (85,000 ha) of cotton area in Australia. Isecticide use has been reduced by about 60% The efficacy has been moderate. The transgenic plants were found to express Bt proteimns only till the 95th day after which fluctuaions were observed. The reduction in expression was primarily due to down regulation which post transcriptional changes of the unstable RNA It was also reported that cotton tannings which increase with growth phase, act as antagonistswith Bt toxins.
- After two years of field cultivation, true resistance to Bt has not yet been detected in Australia.
- Reslistanor to Bt in field populations of Helicoverpa armigera is reported form China
- Annually 100 million dollars are now being spent in the US, only in search of new insecticidal genes.
- In the US, 45% of cotton area this year will be under transgenic cotton (herbicide and insect resistant). However, the area under Bt cotton seems to on the decline.
- Transgenic cotton developed in Australia using the Stunt virus genes, is not performing well.
- Transgenic cotton developed in the US using lectin genes, is not performing well against Heliothis virscens.
- The Australian report that Helicoverpa armigera is at least 100 times less susceptible to Cry toxins as cinoared to Heliothis virescens.
- The Window strategy for insecticide resistance management in Australia is still important for transgenic cotton.
- Cotton Pest spectrum in the US is getting altered after the introduction of Bollgard. Unsprayed Bt cotton sustained 4 times more attack of tarnished bugs, 2.4 times more with boll weevil, 2.8 times more with stink bugs and Spodoptera.
- Due to these changes in pest complex farmers had to spray 3-5 times on boligard as compared to 6-8 times on non-Bt cottons.

Prospects and Potential

The application of biotechnology in cotton farming can be either in the form of production of fermentation products or novel recombinant products for use (example, recombinant Pseudomonas expressing CrylAc) or as transgenic plant with in buitt resistance to biotic and abiotic stresses. Transgenic crops with in built resistance to insect pests and diseases can be extremely useful as this would result in the reduction of insecticide use apart from making pest management simple for the farmer.

Introduction of the bollworm resistant transgenic cotton is expected to reduce the use of chemicals used to control bollw-orm especially Helicoverpa importantly at a time when resistance to most insecticides including pyrethrolids is on the increase all over the world. Even if the introduction of the Bt cotton in India could result in a 25-30% reduction in insecticide use on cotton, this would mean a benefit of about of about Rs 300 crores, apart from the f-avourable impact on the environment so far, transgenic plants have been produced in about 60 plant species.

Cotton has received special attention of the biotechnological companies in the developed countries who were attracted by the profit motives associated with the high value added to the transgenic seeds. The firs Bt transgenic cotton has already been released, as Bollgard. The Delta and Pineland using DP 5410 and DP 5690 as recurrent parents. The D and PL brand BT transgenic were labelled as Nucotn 33B Two million acres were planted in 1996 in mid-south region of USA, mainly in regions where Heliothis virescens was problematic. Nearly 1.2 M acres were under Nucotn 33 B and the rest under Nucotn 35B.

The Bollgard brand Bt cotton seed was sold at US $34-36 as compared to US $8-9 per hectare for non-engineered cotton seed. The average cost of control of cotton insect pests in the US was approximately US $150 in the early 1990s. Hence the prices were still found to be attractive. The transgenics were found to be more effective against Helothis virescens as compared to pectinophora gossypiella and Heliothis zea.

Farmers growing transgenics had to sign'an agreement with D & PL stating that they would not keep seed for planting next year. Quick ELISA tests have been devised to test for Bt toxins in plant parts, to check the illegal spread of transgenics. Two more transgenic cotton varieties tolerant to herbicide Bromoxynil, trademarked as BXN 57 and BXN 53 were developed from the recurrent paren Coker 315 BYTHE Stonevuille pedigree company in collaboration with Calgene Inc. This seed was priced nearly 1.7 times the straight varieties.

In Australia the CSIRO and Narrabri research centre, in collaboration with Monsanto have developed transgenic cotton plarts that were commercial released in 1997. This introduction resulted in 50-60 % reduction in the US $ 93 million spent by farmers each year on insecticides. Pakistan which uses 90% of its total insecticides on cotton alone and China which is one of the major consumers of insecticides on cotton are reportedly strongly considering the prospects of introduction of Bt transgenic cottons.

Cotton acreage in India is only 5% of the total cropped area, yet it consumes more than 50% of the total insecticides used in the country. The pattern of usage, however, is not uniformly, spread, for instance in Andhra Pradesh alone, where cotton cultivation is only 0.3% of the total cropped area of the country, accounts for 17% of the pesticide use on cotton in India. Again within the average picture of the state, coastal AP uses 30% more pesticide than the state average. The approximate estimates of insecticide sales in Guntur district alone, where cotton is grown in 0.15 mHa, is about US $ M.

Clearly areas with maximum pecticides use per hectare, necessitated mostly due to bollworms, are likely targeted market niches for the bollworm resistant transgenic cottons. Moreover, since the price of the transgneic seed in the US is nearly four times the high input cotton cultivators of the irrigated belt Growing Bt transgenics and herbicide tolerant transgenics on marginal land is not recommended by the company as these would not show any impact in the absence of strong insect pressure and broad leaved weeds. In most areas of rainfed cottons the use of insecticides and pest pressures are low hence, it may not be economical in such belts, to encourage the use of transgenics.

Now transgenic production technology for many crops has become an established routine procedure in many countries including India. The research on transgenic cotton in government funded research labs in India is almost in the final stages. Meawhuile Mahyco-Monsanto biotech is ready to market Bt transgenic cotton (bollgard) in India. The introduction of transgenics through these companies are to be in the form of FL seeds wherein Bt genes are inherited from the transgenic exotic germplasm. Such seeds may also be priced at a premium and it remains to be seen how enthusiastic the response of Indian farmers would be to the expensive input. Transgenic releases from government organisations would definitely enforce a competitive pricing to restrict the monopoly of the private companies.

The transgenic technology is also likely to be extremely beneficial to the private sector. For instance, a transgenic with resistance to a specific herbicide marginalises the use of other herbicides for chemical weed control and the chemical company holding the patent on the herbicide might be prove to be beneficial to the resource rich farmers. Clearly, here the targeted market is primarily high input farmers. The monopoly of transgenics due to patenting rights held by the biotech companies may also lead to high pricing of the seeds.

A Significant socioeconomic issue that arises from the introduction of transgenics into the Indian farming system is that the high priced seeds may benefits the prosperous and large farmers thus providing a negative externality on small and marginal farmers. On the other had it is also argued that the developments from the application of biotechnology would be beneficial to low input farming practises wherein the cost of chemical inputs can be minimised. It is now only a matter of time before we experience the full social. economic and environmental impact of transgenics in our country.

Biotechnological Tools for Increasing Productivity of Pulse Crops

Pluses play a significant role in Indian agriculture because they provide protein rich diet to poor people of the country. They contain 20-29% protein i.e. about 2.5 times more than cereals. Inspite of release of large number of high yielding varieties, extension of irrigated areas and use of agrochemical, growth recorded in total production of pulses from 11.82 million tonnes in 1970-71 to 13.19 million tonnes in 1995-96 is considered to be non revolutionary. This resulted in decline in per capita availability of pulses to 37g/person/day.

There are several reasons for stagnation in pulses production such as narrow genetic base, lack of genetic variability in useable genepool, inability to transfer increased input to increase output (seed yield) susceptibility to large number of biotic and biotic stresses, photoperiod sensitive and non synchronous flowering and maturity. It has been estimated that on account of insect pest complex infesting pulse crops alone, nearly 2.0 to 2.5 million tonnes of pulses is lost annually. The monitory value of this loss will be approximately Rs.3000 to 3750 crores. Conventional methods of crop improvement have not improve productivity of pulses with significant pace due to several reasons. These methods are slow, expensive and time consuming. Besides lack of reliable resistant sources, undersirable linkages and long gestation period are considered to be the main limitation of these methods. Hence, there is dire need to introduce innovative biotechnological approaches to enhance pulse productivity to meet out the market demand.

Biotechnology and genetic engineering has emerged as potential tool in recent year for genetic manipulation of crop plants. The said technology promises to improve crop productivity through complementing traditional breeding by decreasing dependence on harmful chemicals pesticides. Fertilizers and antibiotics etc. The tools of biotechnology are now available to plant breeders and can be used effectively to supplement conventional practices in improvement of pulse crops. It is expected that products of biotechnology will be ecologically sage, economically feasible and technologically expectable to Indian farmers.

Effort have been successful to develop several intergeneric and interspecific hybrids using embryo/ovule/ovary/endosperm culture in grain legumes, Several useful genes imparting resistance to various biotic and biotic stresses have been successfully introgressed on wild relatives of lathyrus, lentil, chickpea, pigeonpea urd bean and mung bean. The resistance to Aschochyta blight to chickpea has been successfully transformed from wild species Cicer judaicum to elite chickpea genotype PDG 84-10 at IIPR, Kanpur.

The techniques of in vitro mutagenesis in vitro selection and somaclonal variation are being also successfully employed to various pulse crops to develop promising genotypes resistant to various biotic and a biotic stresses. This technique has resulted success in developing Aschochyta blight resistance in chickpea. Further early results in vitro selection for salinity tolerance in chickpea is also very encouraging at IIPR, Kanpur. Similarly, the technology of cell suspension culture followed by selection of mutants has also been used successfully in major pulse crops at different centres in the country.

Identification and characterization of pathogens and pests is problematic and always pose challenge to plant pathologist and breeders in varietal development of pulse crops. The techniques of molecular diagnosis and DNA finger printing has facilitated resulted easy diagnosis and the precise characterization of pathogeny isolates and insect biotypes at molecular level. This has increased the efficiency of breeding programmes. Important pathogent like Aschochyta rabi and Fusarium oxisporum of chickpea have already been characterized using DNA fingerprinting techniques. In a new world order and changed senario the characterization of germplasm and plant material is very essential to protect out natural wealth. The techniques of DNA fingreprinting have been proved very handy in protecting the intellectual property rights. These techniques are now being successfully used in pulse crops to charac-terize and catalogue the accessions of gene bank to protect the genetic wealth. DNA marker techniques are the new innovations and have tremendous potential in increasing the efficiency of conventional breeding programmes. DNA marker technique can be used for tagging important genes which can subsequently be used for early, easy and precise selection in segregating of promising plants populations. DNA markers have been found linked to wilt resistance in chickpea, powdery mildew resistance in peas and nemarode resistance in mung bean and urd bean.

Transgenic technology which involves insertion of foreign DNA sequence has tremendous potential for improvement of crop plants. Efforts are on to develop transgenic plants resistance to pod borer in chickpea and pigeroxpea using Bt. Crystal protein gene from a soil bacterium. Early results obtained at various centres in the country are very encouraging. Similarly, efforts are also being made to develop transgenic urdbean and mung bean using coat protein gene. Current status and future strategies for genetic manipulation of pulse crops using biotechnological tools will be discussed in detail.

Biotechnological Applications for The Improvement of Indian Tobacco

India is one of the top 10 countries cultivation tobacco for the part ten decades both for domestic and international market, India is the third largest producer of tobacco in the world with an output of approximately 520 m kg. Of which FCV tobacco account for 120 m kg. The economic importance of to Indian exchequer is peogressively increasing in the last 10 years. Tobacco generates 10% of total excise revenue and tobacco 4% of the value of total agricultural exports, India has the unique distinction of growing different types of tobacco in different Agroclimatic zones.

Tobacco provides direct and indirect lively-hood to 26 million in India including 6 million farmers and workers. Many small and marginal farmers were benefited by growing tobacco. Research efforts to identify alternative crops to tobacco indicated that no other crops is as remunerative as tobacco. Stability of the worldwide demand of the crop, the crops hardiness and ability to grow in climate and soils non-suitable for other crops, relative ease in transporting tobacco and tobacco's profitability compated to other crops are incentives for farmers growing tobacco thus making tobacco one of the most appealing cash crops to farmers. Tobacco crops is considered as major employer worldwide. Despite technological progress tobacco cultivation remained as one of the most labour intensive crops among all other agricultural crops. Women in particular and skilled labourers, who would otherwise have little chance of employment are able to earn living by working in tobacco cultivation.

WHO in its report has acknowledged that in developing countries tobacco can create vast employment. In view of its great role in the economy of the country and farmers, further improvement of tobacco crop through Biotechnological tools will benefits them a great deal.

Biotechnological applications: Scientific system of tobacco cultivation in India was introduced with the establishment of Central Tobacco Research Institute (CTRI) at Rajahmundry in 1947. Tobacco research in India over the past 50 years through CTRI has made significant contributions to the economic upliftment of tobacco farmers by identifying high yielding varieties and suitable package of practices This is reflected in the good reputation for Indian Tobacco in the International market, as leaf of desired quality traits as per the specific requirement of traders is produced. Initially, conventional breeding methods were utilised for improving the yield and quality of tobacco. However, later application of biotechnology tools for supplementing the conventional tobacco improvement programmes or initiated. Different biotech techniques like anther culture, fertilized ovule culture micropropagation, protoplast culture somatic hybridization and also genetic engineering are being utilised. These techniques are helpgul is strengthening the conventional breeding programmes where ever they are week and failing to serve the purpose. Progress of biotechnology world, in different aspects, at CTRI is given below:

Yield Improvement

Yield improvement through conventional breeding programmes takes fairly long time. This can be reduced substantially (3-4 years) by culturing anthers (Anther culture). Of elite lines from early breeding populations in tissue culture. Through anther culture several dihaploids are produced and evaluated every year. An improved line D1 was developed, in this way, which is superior in yield. Efforts to incorporate budworm resistance through someclonal variation in this line is in progress. Stable resistant D1 someclones are there in field testing stage.

Resostance to Biotic and Abotic Stresses

Wild species of tobacco possesses resistance to various tobacco pests and disease. It is difficult and in some cases impossible to transfer this resistance from wild species to cultivated tobacco through conventional methods.

In such cases, inter-specific hybrids were rescued through fertilized ovule culture and mass multiplied for further testing by micropropagation, In this way, interspecific hybrids possessing resistance to leaf spot diseases, black shank, root knot nematodes leaf eating caterpillar, aphids, budworm etc. were developed and evaluated for their resistance. Thus, stable pest and disease resistant lines were identified through Biotechnological tools. Efforts to transfer desirable characters from wild species through somatic hybridization is in progress.

Somaclonal variation was attempted to fortigy tobacco lines resistance to white fly (Bemisia tabci Genn.) (leaf curl) and budworm (Heliothis armigera Hb) and the results are encouraging. Somaclonal variation is highly effective to create variation when desired variation is impossible to get through conventional breeding methods and also variation is required only for one character.

Transgenic Bt tobacco cultivars processing resistance to leaf eating caterpillar (Spodoptera litura F) and budworm (Heliothis armigera Hb) were developed. Production of transgenics will give scope for transferring desirable characters from different species and genera.

Preliminary attempts were made to screen lines for drought tolerance in tissue culture. Such efforts will save expenditure on extensive field evaluation and work can be carried out in the off season also. Transgenic tobacco having wheat desiccation tolerant protein were evaluated water stress tolerance.

Herbicide Resistance

Orabache cernua os a root parasite of tobacco causing heavy loss in yield and quality. Glyphosate at lower does was found to kill Orabanche, but it is phytotoxic to tobacco. Hence research on production of herbicide (glyphosate) resistant tobacco lines through tissue culture was initiated. When farmer grows herbicide resistant lines in the field he can use berbicides without any harmful effect on tobacco, for the effective control of weeds and Orabanche. Conservation of germplasm: Germplasm is the valuable source for various characters that are useful in future breeding programmes. This can also be conserved under in vitro conditions for long time. Also lines that cannot be maintained under field conditions due to non-flowering and sterility can be effectively maintained under tissue culture conditions. Interspecific F1s, mammoth mutants, in vitro rescued hybrids protoclones, BT transgenics various mutants are routinely being maintained.

Future Thrusts

Biotechnology can help Indian tobacco farmer in many ways, Particularly in tobacco improvement research and seed purity disputes, if offers a large scope. Research on tobacco improvement benefit farmer as he can get high yielding and pest and disease resistant tobacco for cultivation. In research various aspects like in vitro conservation and molecular characterisation of germplans, isolation, characterisation and transfer of desirable genes from across genera, molecular aided selection in breeding programmes, someclonal variation can play a great role. Utilising Biotechnological techques tobacco lines that can yield higher amounts of phytochemicals which are useful in the production various drugs and pesticides can also be developed.

In recent times many disputes about seed purity among farmers are coming up. At present morphological criteria is only being used for varietal identification. As morphology is highly subjective in nature and also varies with the type of environment the plant is grown, it is becoming highly difficult to solve the disputes. Under this situation molecular characterisation of cultivated varieties will help to identify correct varieties beyond doubt.

In future, giving emphasis on the above said lines will go a long way in benefitting the India tobacco crop as well as the farmer. However, as biotechnology got wider applications, any thing that put the farmer and environment at disadvantage need to be discounrages.

Biotechnological Approaches to Weed Management

Weed management is essential for good quality and quantity of food production. The presence of weeds in general may reduce crop yield by about 30% or more. Being potential pests, weeds have to be removed employing mechanical, chemical cultural and biological means. Weed management through application of chemicals (herbiciedes) is, through effective, not eco-friendly. The herbicides may pollute the environment and many alter the natural equillibrium. They also endanger and alter plant and microbial biodiversity. Effective non-chemical weed management may include various biotechnological approaches using plants and/or their products. These are useful both for the natural equilibrium and the environment. In addition they also condition the soil and improve crop production. Some of the recently evolved biocontrol and biotechnological methods of weed management developed and tried at the National Research Centre for Weed Science, Indian Council of Agricultural Research, Jabalpur are outlined in the following text for ready use for trials and for working our their cost effectiveness.

Management of Phalaris minor in wheat by using Fungi

At present Phalaris minor is one of the most predominant weeds in wheat fields. The Weed spreads through seeds. The weed has developed resistance against popular herbicides like isoproturon. Biotechnological approach using fungi offers an alternative possibility of management of the weed through biocontrol.

(i) Control of Phararis minor using Trichoderma viride: Trichoderma viride is used as a seed dressing fungi in various crops including in wheat for management of seed borne/soil borne pathogens. Interestingly, the fungus has been found to inhibit germination of seeds of Phalaris minor. Application of Trichoderma, viride grown in saw dust and neem oil cake as a base consistently inhibits germination of phalaris minor seeds and reduces vigour of the seedings of the weed. At the same time it improves yield of wheat by about 5%. The fungus has potential of controlling phalaris minor. Soil application of the fungus, wheat seed treatment and spray of the fungus were all effective in controlling the weed.

(ii) Control of Phalaris minor by gilocladium virens: Control of Phalaris minor by ?Gliocladium virens has also been found to have potential for control of Phalaris monor by inhibiting seed germination in a similar manner as does the Trichoderma viride.

Management of Parthenium (Parthenium HysterophorusL) by Marigold (Tagetes PatulaL)

Parthenium is an obnoxious weed of worldwide occurrence, harmful to human and animal health, agriculture, environment and the natural biodiversity. The weed reduces yields of crops and grasses. When shown in parthenium infested field or area, marigold inhibits seed germination and growth of the weed. It also reduces population of the weed. In the next generation, reduction of parthenium population is as height as 85-100%. This has been found primarily due to competition and through all elopathy (affecting parthenium plants by releasing chemical substances including thicophene into soil through roots.)

Built up of the phytotoxic substances of marigold origin prevent parthenium seed germination, growth of plants, and flower and seed production completely in 2-3 years. This technology appears to be economically rewarding for marisgold flowers have good market. This can be achieved simply by spraying marigold seeds over parthenium infested area. The technology is simple, effective, economically rewarding, and ensures self perpetuating control of parthenium in wastelands. Marigold to parthenium plant ratio of 0.5 to 4 appear effective for near complete control of parthenium. Adequate moisture availability in soil facilitates establishment of marigold in parthenium infected area.

Management of Parthenium by fungi: Two fungal species namely Gliocladium virens and Trichoderma viride which have been widely used as seed dressing to present parthogenic damage have also been found effective for parthenium control.

(i) Control of parthenium by Gliocladium virus: Glioladium vieus and 10% neem oil separately as well as in combinations spray control pathenium and other broad leaf weeds like chenopondium album, Melilotus alba and Medicago sp. In wheat crop. The fungus could be explored as a self perp control measure for non-cropped area also.

(ii) Control of parthenium by Trichoderma viride and need oil: Thrichoderma viride and neem oil separately as well as in combinations spray control parthenium under field conditions. The fungi has potential of becoming a self perpetuating means for control of parthenium under non-cropped area. Trichoderma viride also acts as an antidote for parthogens of parthenium that can possibly be used as biocontrol agents. If any of the parthogens attack crops, the Trichoderma virid can be used to overcome the problem.

Management of Parthenium by Utilising as a Source of Nutrients

Parthenium is able to extract nutrients even from nutrient deficient soils in which it grows. It has very height levels of nitrogen (3%), phosphorus (0.2%), potassium (4.5.%) and other macro and micronutrients. Uprooted, dried and cut to small pieces or powdered parthenium plants at preflowering stage can be applied directly to crop fields. Apart from increasing crop growth and yields, the parthenium residue conditions soil with the organic supplement. Keeping safety considerations, it may serve as an incentive for participation of public in its ecofriengly control.

Weed management through mid-summer irrigation and covering the beds with black polythene: Weed problem in most cases is due to seeds of weeds present in soil seed bank which germinate before, with or after planting the frops.

Mid summer irrigation and covering the beds or fields with black polythene thoughout the summer kill most of the readily gemin-able seeds in soil seed bank. Polythene covering can be removed after summer before planting the crops. The treatment may provide near complete weed control. This technique is suitable for areas where summer temperature exceeds 400 c. Weed management through black polythene stripe mulching Where crop is grown in rows, inter row space could be covered with black polythene mulching. This kills weeds appearing in the area due to interception of light necessary for survival of the plants. The are many other ways of biotechnological management of weeds such as bioling water jet treatment of soil, by intensive cropping, by increasing plant density and chosing allelopathic crop species etc.

Biotechnologycal Application

Techniques such as embryo transfer, dilution of semen and its cryo preservation and artificial insemination (A.I.) are well developed and expertise in these fields is available. This technology is extensively used in the case of cows and to some extent Buffaloes. The main lacuna in the case of sheep and goats is the non availability of improved breeds which could be utilised. As these animals are owned by the poor strata of society and there is a mistaken prejudice that they are responsible for environmental degradation no sustained efforts have been made to improve indigenous breeds and thus make available genetic stock for upgrading.

In spite of this the numbers of these animals have been increasing rapidly (*2.5% per year in the case of goats) due to an increasing demand for meat. Our breeds have developed over the centuries to survive and adapts themselves to harsh enviro-nmental conditions but do not universally possess ability to respond to better management and nutrition. Outstanding work in the case of the Sirohi goat was carried out from 1981 to 1993 by the Indo-Swiss Goat Development and Fodder Production Project in Rajasthan where they showed that outstanding animals are available within this breed. However, before the impact of this project could be felt the project was discontinued.

Goats

Out institutes is a very small NGO with only three technically qualified staff, one of who is in an honorary capacity and one on deputation from the Government of Maharashtra. We decided to initially concentrate on crossbreeding using the outstanding South African Boer Goat the worlds only meat goat as an improve breed. With assistance from the Government of Australia we imported 20 embryos of this breed from Australia implanted them in local goats and now have a small flock of Boer goats.

This goat unlike many temperate breeds of Dairy goats have proved to be extremely hardy, producing excellent crossbreed when crossed with local goats. In order to utilise this germplasm effectively we have developed and modified the techniques of semen cryopreservation so that it can be done with low cost equipment producing pellets instead of straws. We have also improvised. A.I,. techniques using cervical insemination to get better conception rates. We have so far trained over 50 trainees among them farmers. We have supplied pure Boer goat frozen semen to Governmental agencies in Orissa, Tamil Nadu as well as to may private goat rearers in Maharashtra.

Our veterinarian is capable of carrying out multiple ovulation and embyo transfer if there is need for rapid expansion of our flock. However, this technology is expensive and only to be used if costs justify its use.

A I technology in goats has been spread to near by villages where fresh diluted semen as well as frozen semen have been used to produce 800 crossbreeds over the past five years. This technology is viable and easily transferable but unfortunately dependent on the sincerity of the inseminator.

Sheep

We were unable to identify a suitable improve breed to be imported which had the most desirable but greatly neglected attribute of prolificacy. We did locate a prolific and unusual breed in the hot humid sundarbands of West Bangal called the Garole. Up to 1993 this breed had remained unnoticed by the community of sheep breeders. We did however import a small flock of Awassi Dairy Sheep from

Israel. This breed has produced an outstanding crossbred animal when crossed with the Malpura of Rajasthan. This crossbred has been developed by the Central Sheep and Wool Research Institute (CSWRI) using Awassi rams obtained from us. The Australian Centre for International Agricultural Research has sponsored a collaborative project between the University of New England, Australia, The national Chemical Laboratory, Pune and the Nimbkar Agricultural Research Institute (A SISTER NGO of Maharashtra Goat and sheep Research & Development Institute), Phaltan entitled," Prolific Worm Resistant Meat Sheep for Maharashtra, India. This project is making full use of Biotechnology. We are comparing three breeds of Indian sheep the Grole, Deccani and the bannur. We hope to evaluate lamb production including prolificacy and resistance to parasites and try and determine the genetic basis of these differences by DNA analysis. We also hope to account for the interaction between measurement of resistance to parasitic nematodes and environmental effects. The project has completed one year and the results so far are encouraging.

Our ultimate objective is to provide shepherds with rams incorporating the prolificacy of the Garole, the growth potential of the Awassi and the meat quality and hardiness of the Bannur. A very important but at present neglected application of Biotechnology which requires considerable investment is in the field of disease diagnosis and vaccine preparation. Once productive animals that respond to inputs are available and the rearers can afford to pay for the above services I am confident that the private sector will enter this field to provide this input. It is imperative to review the cost benefit ratios of Government institutions and their contribution to the Goat and Sheep industries I think you will find that funds given to NGO's are more productive. There is much talk of helping NGO's but real assistance only seems to come from foreign aid agencies which is unfortunate.

Bacterial Biotechnology

The Production of high quality agricultural products in quantities sufficient to feed the world's people is clearly a matter of vital importance. Any general improvement in agricultural practices can have a tremendous economic impact, and the methods of biotechnology suggest at least two important strategies for bringing about widespread improvements. Improvement through the use of symbiotic or potentially pathogenic microorganisms- Plants live in intimate relationship with many microorganisms. Leaf surfaces are covered by layers of symbiotic

& sometimes pathogenic bacterial & yeasts, and the soil close to the root area is populated by bacteria species very different from the species found in soil away from plants. There have been a number of attempts to modify symbiotic microorganisms so as to make them even more beneficial to plants. Improvement through the production of transgenic plants- Plants now being cultivated represent the end result of many years of pain staking efforts at improvement. Traditional plant breeding programs rely on two methods; the selection of advantageous spontaneous mutations & the introduction of desirable traits from closely related species by crossbreeding. Recombinant DNA Technology is bringing revolutionary change to these endeavous.

It enables scientists to select precisely defined genes with wellcharact-erized properties, clone them, and transfer them to a given plant. These genes can be taken from organisms unrelated to the target plant-even from bacteria or animals.Transferring cloned genes from prokaryotes (bacteria) into higher eukaryotes, such as plants, is difficult. However, there are two species of bacteria, Agrobacterium tumefaciens & A rhizogenes, that transfer a small piece of DNA into plants as part of their normal life cycle.

Use of symbionts & Pathogenes

There are many Symbiotic bacterial associated with specific organs of various plants, a simple plant might be to modify such bacterial & then use such interactions with the plants to introduce the modified traits at the appropriate locations. This alternative approach is technically easier, because the engineering of bacterial DNA through recombinant DNA methods is now routine, whereas the manupulation of plant DNA is yet to be perfected.

(a) Protection of plants from frost damage via engineered symbiotic bacteria: One of the earliest examples of successful modification of a symbiotic bacterium was performed in Pseudomonas syringae, which is found at high concentrations on the leaves of many plants. Many strains of this bacterium produce an ice nucleation protein that is apparently located on the surface of the bacteria cell, and the presence of this protein cause the formation of ice at temperature only little below 00 C, inflicting significant frost damage on important crop plants & so facilitating invasion of plant tissues by the bacteria.

(b) Use of nitrogen fixing bacteria to improve crop yields: All animals & plants & most bacteria depend on the availability in their environment of some form of "combined nitrogen" nitrate (No3), or such nitrogen containing organic compounds as amono acids. The huge amount of N2 that exist in the atmosphere are unavailable to the biological world except through the process of nitrogen fixation.

The biological process of nitrogen fixation does not require the consumption of fossil fuels or electricity does not produce environmental pollution. Thus increasing the extent of biological nitrogen fixation has been an important goal for biotechnology. According to one estimate, the biological process fixes times more nitrogen (24 x 10 17 tons/year) than is converted into chemical fertilizers by industrial processes. Thus even a small improvement in exploiting the biological process would have been a global impact. He biological process of nitrogen fixation is complex & consumes a large number of ATP, molecules, because the enzymes involved in it must overcome the same huge activation energy barriers that faces the chemical synthesis of ammonia. Two enzymes are required: component I (nitrogenase) & component II (nitrogenase reductase). After component Ii is reduced by a strong biological reductant, 16 molecules of ATP are hydrolysed to accomplish the reduction of component I. Reduced component II alone is not capable of overcoming the activation energy barrier. Reduced component I finally reduced N2 to two molecules of NH3. Nitrogen fixation is strongly reductive reaction, and the enzymes involved are usually irreversibly inactivated when they are exposed to oxygen.

The species of closiridium & klebsilla fix nitrogen only under angerobic conditions. Other free living baueria, however, can fix nitrogen even under aerobic conditions. The cyanobaueria, which carry out oxygen-evoluing photosynthesis, perform nitrogen fixation only in specialized cells called heterocysts, which do not produce oxygen. Azotobactor consumes oxygen at an extremely high rate, and in doing so apparently protects its nitrogen fixing machinery. Another group of bacteria fix nitrogen only when they are in a symbiotic nitrogen fixer is Rhizobium, which invades the root tissues of teguminous plants, such as alfalfa, pea, clover and soyabean and lives in intracellular vacuoles. The vacuoles become filled with an oxygen-binding protein, leghomoglobin produced by the plants. This creates an environment low in free oxygen where Rhizobium is able to carry out nitrogen fixation.

10

Embryo Rescue and Embryo Culture

Many distant crosses fail due to degeneration of endosperm. In such cases, embryo culture is used to recover hybrid plants; this is called hybrid rescue. Embryo culture, has been widely used for this purpose. Some recent examples are the recovery of hybrids from H. vulgare x Secale creale, T. aestivum x Agropyron repens, H. vulgare x Triticum species, including T. aestivum, etc. Hybrid seeds from T. aestivum X H. vulgare are not obtained, but hybrid seedlings have been obtained by embryo culture. Hybrid embryos from 9-12 day old caryopses were removed and placed on 14-18 day old barley endosperm cultured on a nutrient medium. A high proportion of hybrid embryos developed into seedlings when they were thus cultured.

Embryo-rescue Techniques

Techniques were developed in the early 1900's enabling unripe seed or embryos from adult plants to be rescued to form small plants. This was done mainly with seed which had a very long dormancy period or when the seeds were particularly heterogeneous. With the continuing developments in tissue culture this technique was also used to save embryos from ovules, which had been fertilized but had never developed into viable seed on the mother plant.

Initially, complete ovaries were put in tissue culture whereby seedlings were obtained from embryos which wouldhave died in a later stage of development. Saving embryos that died in an early stage of development came at a later stage, resulting in high-tech ovule- and embryo-culture techniques. Often a combination of these techniques is used: parts of the ovaries are put in tissue culture following which ovule and/or embryo culture is applied.

Embryo Culture

Collecting and conservation of coconut genetic resources are important priorities for breeding programmes (IBPGR 1985). The coconut seednut is characterized by considerable weight and volume. The lack of dormancy renders its transport conditions very difficult and costly, and poses phytosanitary problems.

The use of *in vitro* techniques can facilitate the transport and offer some phytosanitary guarantees. In addition, the cryopreservation of zygotic embryos can also play a major role in the conservation of coconut germplasm and in the exchange of genetic resources.

A summary of the 10 years of research work conducted by Institut de Development de Foret/Department du Plante Oleagineous (IDEFOR/DPO) in collaboration with ORSTOM and CIRAD, for collecting and culturing zygotic embryos is presented.

Collecting Fruits Using in Vitro *Techniques*

Fruit collecting through zygotic embryos involves sampling, disinfection and transferring of embryos to the culture medium. The ideal age of the fruit is about 10-11 months. Embryos of over-ripe nuts may grow prematurely and disinfection becomes difficult at this stage.

Sampling - Sampling consists of cutting the albumen cylinder with the embryo. This operation is done in non-sterile conditions. The nut without the husk is split open. The part with the embryo is placed on the table and a cylinder of albumen around the embryo is cut using a 2.0-cm cork borer. Gouging of the albumen around the embryo serves to protect this organ during storage and disinfection. This piece of albumen is removed just before culturing.

Disinfection - The cylinders are put in a 500-ml flask containing calcium hypochlorite solution (8% active chlorine). During the imbibition, stir the disinfectant in the flask once or twice. After the disinfection, the cylinders can either be stored for subsequent transfer to the growth medium or the embryos can be cultured right away.

Embryo transfer in the field - In the laboratory, the transfer is done in a laminar flow cabinet. In the field, the different operations are conducted inside a field inoculation box equipped with an alcohol lamp for flame-disinfection of the scalpel. The scalpel is heated on the flame, then cooled off inside the jar containing the albumen cylinders with the calcium hypochlorite solution. Using the scalpel, the albumen core or cylinder surrounding the embryo is carefully cut to isolate the embryo which is then placed in a sterile petri dish, washed with distilled water, and placed in a sterilized 30-ml flask containing the prepared culture medium. The tubes of cultured embryos can either be put directly inside a culture room (at 25-27°C), or conserved under a shade before the transfer to the culture room. If a tissue/embryo culture laboratory is near the collecting site, the albumen cylinders can be sampled in the field and brought to the laboratory. The cylinders should then be put temporarily in a KCl solution, before the excision of the embryos.

Storage of Albumen Cylinder

After the disinfection, the albumen cylinder with the embryo is transferred into a 30-ml flask of storage solution containing 16 g/L KCl. The scalpel should be regularly disinfected by heating. The different operations are conducted over a flame to avoid any contamination by microorganisms. The cylinder with the embryo can be transferred this way to the laboratory for culturing. After the storage period, maximum 14 days, the cylinder is disinfected with calcium hypochlorite solution for 20 minutes. After the excision, the embryo is then washed with sterile distilled water and put in the culture medium. Direct culturing in the field gives a higher contamination rate (10%) than culturing after storage of the cylinders (5%). Both methods are acceptable since their contamination rates vary only slightly from that of the control cultured in the laboratory. The latter has a contamination rate of 3-8%.

Culture Medium and Condition Applied

A very simple culture medium that ensures adequate development of the embryo *in vitro* and enhances its subsequent growth in natural condition can be used. This medium is composed of the mineral solution of Murashige and Skoog (1962) enriched with 41 mg/L of iron-EDTA, 100 mg/L of sodium ascorbate, 60 g/L of sucrose, 2 g/L of active charcoal and 7 g/L of agar at pH 5.5. The culture is placed in the dark at 27 ± 1 °C until the appearance of the plumule. It is then transferred to another room and cultured under the light (35EMm-2S-1) with a 12-hr photoperiod. Subculturing is done every month.

Acclimatization

One month before the transfer into the nursery, the *in vitro* plantlets are subcultured on a medium with a higher dose of sucrose (60, 90, 120 g/L). After 1 month in this medium, the plantlets are washed with tap water and transferred by batches to sand which was previously sterilized by autoclaving. The use of a plastic cover during the first 2 weeks allows a saturated moisture condition. The plantlets in the batches are watered regularly (3 times/week) to keep the sand moist for 1 month and then they are uncovered and exposed in the open air. A nutritive solution is added to the batch every 2 days. After 3 months, plantlets are transferred from the sandy batches to polybags with compost. The nutritive solution is regularly applied to the plantlets to stabilize growth.

Haploid Production

Haploids are defined as saprophytes with gametophytic chromosome number and have been produced in a variety of plant species using a variety of methods. Although, the significance of haploids in genetics and plant breeding has been recognized for long time, with the advent of new biotechnology it has received renewed emphasis, so that the production of haploids has become an important component of biotechnology programmes in different countries.

Although, haploids could be produced following delayed pollination, irradiation of pollen, temperature shocks, colchicine treatment and distant hybridization, the most important methods currently being utilized under biotechnology programmes include

(i) anther or pollen culture and ovule culture and

(ii) chromosome elimination following interspecific hybridization (bulbosum technique).

Factors Affecting Haploid Production - A number of factors influence androgenesis in vitro. The genotype of the donor plant plays a significant role in determining the frequency of pollen plant production. Anther wall factors also support pollen embryo development.

Histological studies support this view - As induction of the pollen into embryoids occurs most easily within the confines of an anther, the anther wall seems to provide a nursing effect. There are two schools of thought regarding the role of the anther wall. One is that it may have a stimulatory effect on the growth of pollen embryos (probably due to the presence of enhanced levels of some amino acids such as glutamine and serine); the other view holds that it may emanate some inhibitory substances into the culture medium thereby blocking the growth of more pollen into embryos. The culture medium also plays a vital role since the requirements vary with the genotype and probably the age of the anther as well as conditions under which donor plants are grown. The medium should contain the correct amount and proportion of inorganic nutrients to satisfy the nutritional as well as physiological needs of the many plant cells in culture.

In addition to basal salts and vitamins, hormones in the medium are critical factors for embryo or callus formation. Cytokinins (e.g. kinetin) are necessary for induction of pollen embryos in many species of Solanaceae except tobacco. Auxins, in particular 2,4-0, greatly promote the formation of pollen callus in cereals. For regeneration of plants from pollen calli, a cytokinin and lower concentration of auxin are often necessary.

Sucrose has been considered the most effective carbohydrate source which cannot be substituted by other disaccharides. Glucose can be used in anther culture in some cases but fructose is far less effective. The concentration of sucrose also plays an important role in induction of pollen plants. Activated charcoal is also added to the culture medium.

It helps in the removal of inhibitors from the agar used for gelling the medium. Another role assigned to activated charcoal is the adsorption of 5-hydroxymethylfurfural, a product of sucrose dehydration during autoclaving, assumed to bean inhibitor of growth in anther cultures. Certain organic supplements added to the culture medium often enhance the growth of anther cultures. Some of these include the hydrolyzed products of proteins such as casein (found in milk), nucleic acids, and others. Coconut milk obtained from tender coconuts is often added to tissue culture media. It contains a complex mixture of nucleic acids, sugars, growth hormones and some vitamins. The physiological state of the parent plant plays a role in haploid production. Success in haploid induction is in part dependent on knowledge of the physiology of the pollen yielding plant. In various plant species it has been shown that the frequency of androgenesis is higher in anthers harvested at the beginning of the flowering period and declines with plant age.

This may be due to deterioration in the general condition of the plants, especially during seed set. The lower frequency of induction of haploids in anthers taken from older plants may also be associated with a decline in pollen viability. Seasonal variations, physical treatment, and application of hormones and salts to the plant also alter its physiological status, which is reflected in a change in anther response. Temperature and light are two physical factors which play an important role in culture of anthers. Higher temperatures (30°C) yield better results. Temperature shocks also enhance the induction frequency of microspore androgenesis. Frequency of haploid formation and growth of plantlets are generally better in light.

Certain physical and chemical treatments given to flower, buds or anthers prior to culture, can be highly conducive to the development of pollen into plants. The most significant is cold treatment.

The developmental stage of pollen greatly influences the fate of the microspore, Androgenesis occurs when a microspore or pollen is induced to shift from a gametophytic pathway to a sporophytic pathway of embryo formation. Anthers of some species (Datura, tobacco) give the best response if pollen is cultured at first. mitosis or later stages (postmitotic), whereas in most others (barley, wheat, rice) anthers are most productive when cultured at the uninucleate microspore stage (premitotic). Anthers at a very young stage (containing microspore mother cells m tetrads) or a late stage (containing binucleate, starch filled pollen) of development are generally ineffective, albeit some exceptions are known.

Differentiation of Pollen - An anther or pollen has two distinct morphologies which have the potential to develop into two different types of generations gametophyte and sporophyte. This is referred to as pollen dimorphism.

The pollen capable of forming haploids (embryogenic pollen) is small with a cytoplasm that stains less intensely with acetocarmine than does gametophytic pollen. Such pollen is usually low in quantity and accounts for a very low frequency of haploid formation. Exposure of plants to low temperatures before flowering increases the frequency of embryogenic pollen.

The pollen mother cell is diploid and has a determinate structure. It undergoes cell reorganization for differentiating into microspore (haploid) cells which are indeterminate structures. Determination into either androgenic (gametophytic) or embryogenic (sporophytic) grains is possibly a function of differential gene activity. Elaborate endoplasmic reticulum, abundance of ribosomes, and normal mitochondria are usual manifestations of gametophytic pollen, whereas quiescence and repression of organelles in micro spores lead to differentiation of embryogenic pollen. Gametophytic pollen germinates in the presence of metabolic inhibitors in contrast to embryogenic pollen; which does not germinate because of sensitivity to these antimetabolites.

Development of Androgenic Haploids - Under normal conditions the microspores mature into pollen grains in situ but in culture their morphogenesis is altered. Depending on the composition of the medium, development may lead either to the formation of embryoids and/or plantlets or a mass of parenchymatous callus. Four pathways based on the few initial divisions in the microspores have been identified as leading to in vitro androgenesis. In pathway the microspores divide by an equal division and two identical daughter cells contribute to sporophyte development. Vegetative and generative cells are not distinctly formed in this pathway (e.g. Datura in noxia). In pathway 2 the division of uninucleate microspores is unequal, resulting in the formation of a vegetative and a generative cell.

The sporophyte arises through further divisions in the vegetative cell while the generative cell either does not divide or does so once or twice before degenerating (e.g. Nicotiana tabacum, Hordeum vulgare, Triticum aestivum, Triticale spp., and Capsicum annuum). In pathway 3 the uninucleate microspore undergoes a normal unequal division but the pollen embryos are predominantly formed from the generative cell alone (e.g. Hyoscyamus niger).

The generative cell either does not divide at all or does so only to a limited extent. In pathway 4 the division of microspore is asymmetrical as in pathway 2. Both vegetative and generative cells divide further and contribute to development of the sporophyte (e.g. Datura metel, Atropa belladonna). Irrespective of the above early pattern of microspore divisions, the embryogenic pollen grains ultimately become multicellular and burst open, gradually assuming the form of a globular embryo. This is followed by the normal stages of postglobular embryogeny until development of a plant. Alternatively, the multicellular mass liberated from the bursting pollen grain proliferates to form callus which may later differentiate whole plants either on the same medium or on a modified medium. It is sometimes possible to obtain and rogenic haploids via embryo formation or through callusing within the same species (Oryza sativa) by manipulating constituents of a medium.

Protocol for Pollen Culture

1. Flower buds are collected from the plants after all preparations for culturing in the laboratory have been made.
2. Buds are collected in a nonsterile petri dish and the length of each bud measured using a cm scale. Buds of corolla length 21-23 mm are chosen as the pollen in them would have usually completed the first mitosis.

3. The buds are chilled for 12 days (7 to 8°C) in a refrigeration unit.
4. They are then surface sterilized in a petri dish containing a suitable sterilizing agent, for instance 0.01 % solution of mercuric chloride.
5. Using sterile, double distilled water, the buds are rinsed 3-4 times in a sterile air cabinet.
6. Using forceps and a dissecting needle, the buds are carefully teased open and the anthers removed.
7. The anthers dissected out from each bud are grouped separately on removal.
8. One anther from each group is then removed and squashed in acetocarmine stain in order to determine the stage of pollen development. The pollen should be in the first pollen division.
9. For each culture, anthers from tobacco buds are placed in 5 ml of liquid medium in a petri dish.
10. After regular intervals of 6, 10 and 14 days of culture, the anthers are removed from the culture and discarded. That ensures the dehiscence of all anthers and release of pollen into the culture medium.
11. The dishes are then sealed with parafilm and incubated at 28°C in darkness for the first 14 days of culture.
12. After 14 days, the cultures are transferred to an illuminated growth chamber (500 lux, 12-h daylength, 25°C).
13. The growth of haploid embryos to plantlets developed from released pollen should be observed.
14. Follow steps 13 and 14 given in the protocol for anther culture for growing the plantlets to maturity.

Sometimes the nurse culture technique is used for raising haploid tissue clones from isolated pollen grains. Intact microspores are mounted on a filter paper disk which is then placed over an intact anther or callus derived from the floral parts. The extract of cultured anthers has equal nursing effects on the growth and development of microspores and may be used in place of a nurse callus or anther. A purely synthetic medium developed for raising haploid plants from isolated pollen grains of tobacco and Datura may also be suitable for other systems. The efficiency of isolated pollen cultures for haploid production may be increased by the gradient centrifugation technique. Anthers at the proper stage of development are gently macerated to obtain a microspore suspension. After removing the debris either by sieving or filtration, the suspension is layered onto a sucrose solution (30% or 55% percoll and 4% sucrose) and centrifuged at 1200 g for 5 min. The pollen (vacuolated and starch filled) with embryogenic potential forms a band at the top of the sucrose solution while non embryogenic pollen (starch-filled) settles to the bottom.

The upper band containing embryogenic or sporophytic pollen is now pipetted out and suspended in the washing medium, centrifuged again, and the supernatant removed. Finally, the pellet containing these grains is resuspended in the culture medium. Such a pollen suspension gives a better response than control in terms of induction of cell division and continued growth of the callus.

Haploids from Unpollinated Ovaries - Production of a haploid individual by development of an unfertilized egg cell as a result of delayed pollination is referred to as gynogenesis. Gyno genesis is found in the interspecific cross Solanum tuberosum (2n = 4x) x S. phureja (2n = 2x), resulting in the production of a dihaploid (2n = 2x) potato. San Noeum (1976) was the first to demonstrate that gynogenesis, an essentially in vivo phenomenon, can be induced under in vitro conditions. She obtained gynogenic haploids using an ovary culture of Hordeum vulgare.

Subsequently, gynogenic haploids have been obtained from unpollinated ovaries and ovules. Induction of haploids from female gametophytes may not be inaccessible. Diploidization of Haploids Plants - Haploid plants obtained either from anther or ovule culture may grow normally under in vitro conditions up to the flowering stage but viable gametes are not formed due to the absence of one set of homologous chromosomes and, consequently, there is no seed set. The only mechanism for perpetuating the haploids is by duplicating the chromosome complement in order to obtain homozygous diploids. In pollen derived plants duplication of chromosomes may occur spontaneously in cultures but due to the small percentage of such double-haploids it is necessary to diploidize the haploids by chemical means. A simple procedure designed to achieve diploidization involves immersion of very young haploids in a filter sterilized solution of colchicine (0.4%) for 2-4 days, followed by their transfer to the culture medium for further growth. In this procedure chromosome or gene instabilities are minimal compared to other methods of colchicine or chemical treatment. Uses of Haploids - As a result of haploid induction followed by chromosome doubling, homozygosity can be achieved in the quickest possible way making genetic and breeding research much easier. The genetic segregation is simplified in homozygotes, recessive genes not being masked by dominant ones. Homozygosity is still more important for those plants which have a very long juvenile phase, such as fruit trees, bulbous plants, and forest trees. Even though, homozygosity can be achieved through self pollination, it is a long process in such plants. As a result of homozygosity, lines which produce pure F1 hybrids are made possible.

When working with plants that are normally polyploid, it is very useful by haploid induction to work at a low ploidy level. Monoploids have an advantage for the mutation breeder in that recessive mutations are immediately discernible. By haploid induction followed by chromosome doubling it is possible to obtain exclusively male plants. An important example of this is Asparagus officianalis in which male plants have a higher productivity and yield earlier in the season than female plants. If haploids are produced from anthers of male Asparagus plants, these are either "X" or "V"; chromosome doubling of V results in supermale plants VV which can subsequently be vegetatively propagated. If XX is crossed with VY, XY plants result.

When homozyotes are artificially made into diploids in vivo with the use of colchicine, problems may arise which are not usually found in vitro. In vitro haploids often double spontaneously, giving rise directly to homozygotes. It is much easier to work with haploid protoplasts rather than diploid ones for somatic hybridization. Problems Concerned with Haploid Production - Although haploids have been raised mainly from anther culture of a large number of species, this technique has not proved successful in respect of all genotypes of crop species. A number of problems are encountered with the induction of haploids.

1. Anthers often fail to grow in vitro or the initial growth is followed by abortion of the embryos.
2. The tissue or callus developing from the anther generally comprises a chimera of diploid, tetraploid, and haploid cells.
3. Selective cell division must take place in the haploid microspores, con comitantly restricting proliferation of unwanted diploid and polyploid tissues. Selective cell division is often impossible.
4. Formation of albinos in anther cultures, especially with cereals can neither be avoided nor the loss of plants due to albinism reduced.

5. The technique of inducing haploids in vitro is not economically viable due to low success rate.
6. Callus derived from anther or pollen in a medium supplemented with growth regulators is usually detrimental for haploid production.
7. It is difficult to isolate a haploid from a mixture of haploids and higher ploidy levels since the polyploids outgrow the haploids.
8. Doubling of haploids is time consuming and may not always result in the production of a homozygote. Double haploids sometimes exhibit segregation in their progeny.
9. The anther culture technique works well on plants belonging to only a few families, of which Solanaceae is one.
10. The frequency of embryoids obtained is often very low.

Haploid Plants Raised Through In Vitro Gynogenesis

Family	***Plant Species***
Chenopodiaceae	Beta vulgaris
Asteraceae	Gerbera jamesonii Helianthus annuus
Poaceae	Hevea brasiliensis
Liliaceae	Coix lachryma-jobi
	Hordeum vulgare
	Oryza sativa
	Triticum aestivum
	Zea mays
	Lilium tuberosum
L. davidii Saliacea	Populus X Simonigra
Solanaceae	Nicotiana tabacum
	N. rustica
	Petunia axillaris
	Solanum tuberosum

Monoploid Production Through Chromosome Elimination - Here monoploids are defined as sporophytes with the basic gametic chromosome number. They have a single genome or chromosome set ($2n = X = 7$ in Hordeum vulgare). Monoploids are generally sterile but by doubling the chromosomes, homozygous diploid fertile plants are produced. Monoploids can be produced by interspecific hybridization followed by chromosome elimination.

The hybridization method consists of crossing cultivated barley, Hordeum vulgare (2n = 2X = 14), With the wild, diploid cross pollinating and perennial H. bulbosum (2n :::: 2X = 14). This technique, the Bulbosum method, consists of the following steps the female gamete of barley is fertilized by the H. bulbosum gamete. During formation of the embryo the chromosomes of H. bulbosum having the barley genome in the embryos are eliminated. These are cultured in vitro as immature embryos. Plantlets from these monoploid embryos can be made to give fertile flowers bearing homozygous offspring, following an efficient chromosome doubling technique.

There are two routes by which monoploids can be induced artificially. One route is based on the male gamete (microspore) and the other on the female gamete (megaspore). Potentially, the first route, via anther or microspore culture, has the advantage because there are far more potential monoploids per spike in the form of male gametophytes than female gametophytes. Both methods are based on embryogenesis and the development of plants from monoploid embryos, followed by chromosome doubling to obtain homozygous diploids.

The advantage of monoploids as tools in plant breeding or genetics becomes more apparent when their direct application is visualized:

1) They provide the quickest possible way to complete homozygosis.
2) They may serve to recover recessives
3) Linkage data can be obtained directly by sampling gametes as monoploids.
4) Doubled monoploids give an immediate product of stable recombinants from species crosses.
5) Monoploids can be used to determine homology within a genome and between genomes.
6) They are ideal for the study of mutation frequencies and spectra.
7) They provide an ideal system for fundamental cell biological problems (i.e., biosynthesis) in cell and protoplast culture.
8) Monoploid cells as protoplasts provide unique material for gene transfer, host pathogen reactions, and cytoplasmic and/ or chromosomal incompatibility.
9) For breeding purposes one of the main advantages of using monoploids is that completely homozygous lines are produced directly from gametes of F1 hybrids or from later (advanced) selections. This allows for a direct fixation of quantitative characters.

10) For practical plant breeders, this saves time and the desired product is readily recognizable.
11) Monoploid protoplasts are a powerful tool in plant modification and somatic hybridization.

Anther and Microscope Culture - One of the very popular methods for production of haploids is through culturing anthers or microspores on artificial culture medium. This leads to the growth of microspores into saprophytes. After the initial reports of successful production of haploids from anther culture in Datura, haploids have been obtained in more than 150 species belonging to 23 families of angiosperms. These include a wide variety of economically important species.

More often, anthers rather than microspores are cultured, since the extraction and culture methods for microspores differ and have been successful only in a few species (Datura inoxia, Nicotiana sylvestris, N. tabacum, Oryza sativa, etc.).

The following parameters have been recognized as particularly important for successful anther and microscope culture;

(i) conditions of growth of donor plant;
(ii) genotype of donor plant;
(iii) the pretreatment;
(iv) the developmental stage of anther/microspore;
(v) the culture medium and the conditions during culture growth.

In either case, flower buds are brought to a laminar flow chamber and sterilized using appropriate chemical treatment before their dissection. While removing anthers from the flower buds, care is exercised to avoid injury, because injury leads to development of callus, giving a mixture of diploids, haploids and aneuploids.

The anthers are generally cultured on solid agar medium, where they may directly give rise to embryoids or may lead to callus formation, before differentiation. The embryoids develop into haploid plantlets, which are colchicine treated to get diploid homozygous plants to be field tested for selection.

Microspore culture may be preferred over anther culture, even though the degree of success is low in this case. Microspores are collected first using the following steps:

(i) anthers, dissected as above, are taken in a sterile beaker containing liquid medium, and are pressed with a glass rod or syringe piston to allow the microspores squeeze out;

(ii) the suspension with anthers and microspores is filtered through a nylon sieve, which allows only the microspores to pass through.

(iii) the filtrate is centrifuged thrice, for 5 minutes each, at 500-800 rpm (the pellet resuspended in a fresh medium each time);

(iv) the microspores are inoculated on a solid or in a liquid medium maintained at 25°C and 16/8 hr photoperiod.

The microspores may develop directly into embryoids within 15 days or follow one of the several indirect paths to produce haploid plantlets. In anther culture as well as in micros pore culture, spontaneous doubled haploids (SDH) are also obtained, so that no colchicine treatment will be needed in such cases. Efforts are being made to increase the frequency of SDH, so that these can be multiplied and directly field tested.

Ovule Culture - Haploid have also been successfully produced from cultured female gametophytes. Success in this connection was first achieved with some Gymnosperms like Zamia, Ephedra and some Cycads. Although initial attempts for the production of haploids from cultured ovules in an giosperms were unsuccessful, recently haploids could be obtained from cultured ovaries in a few crop species, like barley, wheat and tobacco. With more efforts in the future, haploids from ovule culture will be produced in a number of plant species. Further research is, therefore, necessary to improve our understanding about the following;

(i) the events leading to induction of haploidy in female gametophytes,

(ii) the factors which control the in vitro development of proembryo into the fully organized plants and

(iii) the differences in the growth patterns of in vitro development of unfertilized female gametophytes (in ovules) and the male gametophytes (in pollen).

Barley Haploids Through Crosses with Hordeum Bulbosum - A situation analogous to somatic reduction occurs when in a interspecific or intergeneric hybrid, chromosomes of one parental genome are preferentially eliminated. This method has been extensively utilized for production of haploids in barley and also in wheat. The production of haploids in barley in this manner involves interspecific hybridization of Hordeum vulgare (2n = 14) with H. bulbosum (2n = 14), a cross pollinated, self incompatible perennial species. Most progeny (95%) are barley haploids, remainder being diploid hybrids. A cross between 2x barley and 4x H. bulbosum mainly produces triploid hybrids.

Cytological examination of early embryo and endosperm in VV X BB or. BB X VV hybrids revealed gradual elimination of chromosomes. While most genotypes of barley readily produce haploids, chromosome elimination seems to be controlled by genotypes of both parents.

Kasha and Kao (1970) summarized the findings for haploid production using the above technique as follows:

(i) a very high frequency of haploids can be obtained and these haploids are not associated with any diploid hybrids;

(ii) these haploids can be obtained with any cultivated barley and would resemble cultivated barley genome in reciprocal crosses. Contrary to these observations, Lange (1971) showed that haploids obtained due to chromosome elimination are associated with hybrids.

With embryo culture technique, he was able to obtain viable hybrids from several cross combinations, although frequency of hybrids was low except in case of VBB hybrids, which were quite frequent.

Wheat Haploids Through Crosses with Hordeum H.Bulbosum - Barclay (1975) demonstrated that the haploids in hexaploid wheat (var. Chinese Spring) can also be produced through chromosome elimination in hybrids of wheat with H. bulbosum (both 2x and 4x). A frequency of 13.7% grain set with 2x bulbosum and 43.7% grain set with 4x bulbosum, were obtained. In order to explore the possibility of further use of this technique in wheat breeding programmes, studies were conducted at Cambridge (U.K.) on the crossability of wheat with H. bulbosum. It was found that most of the European varieties did not cross, but four of the six Australian varieties were successfully crossed.

A similar crossability pattern of wheat varieties was observed with rye also so that it was concluded that crossability with rye and H. bulbosum was controlled by same genetic factors. Through the use of chromosome substitution lines, it was also shown that 1B, 1D, 2D, 3A, 3D, 4A, 4B and 6D in the variety Hope have genes promoting crossability, while 5A and 5B have genes reducing crossability.

In view of the limitation of crossability with only few wheat varieties, H. bulbosum can not be extensively used for production of haploids in common wheat.

Haploids of Other Hordeum Species from Interspecific Hybrids - Chromosome elimination is also quite common among other Hordeum interspecific hybrids not involving H. vulgare, and has been reported in 19 such hybrids. For instance, H. bulbosum (4x)x H. secalinum (4x) gave haploids of H. bulbosum, while H. secalinum (4x)x H. bulbosum (2x and 4x) and H. secalinum (4x) x . H. vulgare (2x) gave haploids of H. secalinum along with hybrids.

Wheat Haploids From Wheat x Maize Crosses M.D. Bennett and his coworkers in Cambridge, U.K. (now at Kew, U.K.) demonstiated that wheat x maize crosses can be made successfully, and hybrid embryos can be rescued. Due to elimination of maize chromosomes, wheat haploids can be obtained.

Mechanism of Chromosome Elimination - The Mechanism of chromosome elimination in Hordeum hybrids was studied by N.C. Subrahmanyam and K.J. Kasha (1973) and by M.D. Bennett and his associates (1976). The following conclusions were made:

(i) normal double fertilization occurs in interspecific crosses as confirmed through cytological study;

(ii) after fertilization, there is a gradual and selective elimination of H. bulbosum chromosomes from nuclei of endosperm as well as embryo cells, so that eventually haploid embryos are produced. No evidence of the elimination of H. vulgare chromosomes was available.

Differences were, however, observed in the rates of chromosome elimination and these were attributed to genotypic differences in the stocks used in hybridization programmes.

In 1970's, M.D Bennett and his coworkers, then at Combridge (U .K), had examined the possible causes of chromosome elimination and concluded that sudden shortage of proteins in the developing embryo and endosperm, and the better ability of vulgare chromosomes to form spindle attachments relative to bulbosum chromosomes may be responsible for elimination of the bulbosum chromosomes.

Other possible causes like differences in mitotic cycle, congression during mitosis, etc. were ruled out by them.

It has also been demonstrated that the elimination of bulbosum chromosomes is under genetic control. 80 and Kasha (1975) used primary trisomics and monotelotirsomics in crosses with tetraploid H. bulbosum and concluded that both arms of chromosome 2 and the short arm of chromosome 3 of H. vulgare are responsible for chromosome elimination although their effect may be neutralized or offset if sufficient dose of bulbosum chromosomes is available.

The effect of these vulgare chromosomes, however, will be indirect. Detection and Identification of Haploids -A number of methods have been developed for the detection of haploids in a population of diploids. Although the most effective method is the use of genetic markers to be discussed in the next section, phenotypic effects can also be used for identification of haploids.

The vegetative and floral parts and the cell size in haploids are reduced relative to diploids. In maize, Greenblatt et al (1967) selected 45 seedlings out of 10,000 seedlings, on the basis of size (i.e. thin coleoptiles, etc.) and five of them were found to be monoploids showing some success in screening for haploidy.

Stomata size, pollen size and pollen abortion can be other criteria to help initial screening, although in each case cytological confirmation will be essential.

Genetic Markers - Many genetic marker systems are used for recognizing the haploids. The basic principle involved is that the hybrids and nonhybrids should differ, so that any failure of fertilization and parthenogenetic development of embryo can be recognized in the progeny at the seed stage, seedling stage or adult plant stage.

For instance, if a female parent in a cross carries a recessive marker and is pollinated by dominant stock, then any recessive progeny would be generally a maternal haploid, although possibility also exists of its being a diploid of one of the following types:

(i) diploid hybrids, with mutation or deficiency for the marker gene in male parent,

(ii) diploid hybrids with suppression of dominant marker, due to a suppressor gene or due to a disease,

(iii) diploids which are maternal due to apomictic development.

If the recognition needs to be done at the seed stage, then the marker should express itself in the late stage of embryo development, such as aleurone colour marker. If screening is to be done at seedling stage, then the marker should clearly express in the seedling (e.g. the purple seedling colour). This technique of genetic markers has been extensively used in developing maize monoploids. Most extensively used markers include a 1 for brown coloured aleurone, plant and root and 1g for liguleless, their corresponding dominant markers being 'A' for purple colour and '1g' for liguled character. If the female parent carries 'a' and/or 1g, then the monoploids will be recognized by lack of purple colour and/or ligule, in crosses with male parent carrying

'A' and/or '1g' Another critical marker gene is 'R' allele (Rnj-cudu), which produces deep pigmentation of aleurone, and thus allows screening at the dormant seed stage. Other markers included purple plumule (A, Put, Pu2), scutetllum colour (A, C, R genes, and S1, S2 and 53 genes), and purple coleorrhiza (Pc1, PC2, PC3 and PC4).

These markers did not prove to be equally useful, because the frequency of these markers in maize populations is otherwise also quite high. Endosperm markers, however, are not as useful because very often monoploid embryo in the seed is associated with triploid endosperm, SQ that the latter will exhibit the male dominant endosperm trait. It should also be noted that for the recognition of paternal or androgenetic haploids, the dominant marker should be present in the female parent so that the androgenetic haploids may be phenotypically different from the hybrid.

In most of the above cases when genetic markers are used, 90-98% of the progeny can be discarded rather early and a very high frequency of haploids is recovered in the remaining population. The above technique of using genetic markers is particularly useful for the development of inbred lines in maize, but in crops like barley and wheat which are self pollinated and where doubled haploids need to be used for cultivation, the availability and utility of genetic markers may be limited.

Hybrid Sorting in Haploid Breeding - One of the essential steps in haploid breeding involves selection of superior plants among haploids derived from F1 hybrids through anther culture. It is popularly described as hybrid sorting and virtually means selection of recombinant superior gametes. The haploid method of breeding involving hybrid sorting is considered superior over pedigree and bulk methods, firstly because the frequency of superior gametes is higher than the frequency of corresponding superior plants in F2 generation, and secondly because haploid breeding reduces significantly the time required for development of a new variety.

For instance, if we assume that the .frequency of superior F1 gametes is one in one hundred, then the frequency of homozygous F2 plants derived from the fusion of two such superior similar gametes would be one in ten thousand. Therefore, smaller populations of doubled haploids derived from haploids will need to be handled. The above technique of hybrid sorting has been successfully utilized, particularly in China and Japan, for development of new varieties in several crops.

In Japan, an excellent tobacco cultivar, 'F-211' bred through anther culture, is resistant to bacterial wilt and has a mild smoking quality. In China, more than 100 rice varieties developed using the same technique, gave an increased yield and, therefore, occupy significant areas of rice cultivation. Haploids have also been successfully utilized for breeding of barley, maize, sugarcane, oilseed rape or canola (B. napus), bell peppers and some other crops.

In barley, atleast two varieties (Mingo, Gwylan) have been produced in Canada using haploids derived through bulbosum technique of chromosome elimination. The varieties were superior for characters like disease resistance (against powdery mildew and barley yellow mosaic virus) and other yield contributing traits. Similarly, in maize superior inbred lines were developed through screening of haploids derived from anther culture in a hybrid. In sugarcane, selection among anther-culture derived haploids led to the development of superior lines with tall stem and higher sugar content.

In Brassica napus also, anther derived doubled haploids, produced in Canada, had low erucic acid and glucosinolate content. In another case, a haploid derived from twin seedlings in B. napus led to the development of a superior cultivar 'Maris Haplona'. In bell peppers, DH lines produced from anther derived haploids, exhibited all shades of colour ranging from dark green found in one parent to light green found in the other parent of the hybrid employed.

Bibliography

Arthur W. Gilbert, Mortier F. Barrus and Daniel Dean: *Growing and Breeding of Potatoes*, Asiatic Pub, Delhi, 2006.

Bag, A. K.: *Mathematics in Ancient and Medieval India*, Varanasi: Chaukhambha Orientalia, 1979.

Boyer , J.S.: *Measuring the Water Status of Plants and Soils,* Academic Press, N.Y., 1995.

C.R. Sudharmai Devi: *Analytical Procedures in Soil Science and Agricultural Chemistry*, Agrotech, Delhi, 2004.

Crump, Thomas: *The Anthropology of Numbers*, Cambridge, Cambridge University Press, 1990.

Daniells J.: *Illustrated Guide to the Identification of Banana varieties in the South Pacific*, Canberra, ACIAR, 1995.

Diederichsen A.: *Coriander (Coriandrum sativum L.).* Rome, International Plant Genetic Resources Institute, 1996.

Featherly H. I.: *Taxonomic Terminology of the Higher Plants*, USA, Iowa State College Press, 1954.

Ferentinos L.: *Proceeding of the Sustainable Taro Culture for the Pacific Conference*, Honolulu, HITAHR, 1993.

G.L. Maliwal and L.L. Somani: *Nature, Properties and Management of Saline and Alkali Soils*, Agrotech, Delhi, 2010.

Hardy B.: *Biology and Agronomy of Forage Arachis*, Cali, International Centre for Tropical Agriculture, 1994.

I.D. Tyagi: *Plant Breeding and Genetics at a Glance*, South Asian, Delhi, 2005.

J S Amarnath and A P V Samvel: *Agri-Business Management*, Satish Serial Pub, Delhi, 2008.

Jacquat Christiane: *Plants from the Markets of Thailand*, Bangkok, Duang Kamol, 1990.

Kim J. S.: *Sources of Resistance to Viruses of Pepper (Capsicum spp.): a Catalogue*, Taipei, AVRDC, 1994.

Lorenzen, S.: *The Phylogenetic Systematics of Freeliving Nematodes*, London, The Ray Society, 1994.

M.S. Mondal: *Pollen Morphology and Systematic Relationship of the Family Polygonaceae*, BSI, Delhi, 1997.

Matsuoka H.: *Cultivation of Panicum Genetic Resources for Evaluation of Characteristics*, Tokyo, JICA, 1997.

N. Kumar: *Breeding of Horticultural Crops : Principles and Practices*, New India Pub, Delhi, 2006.

Nobel, P. S.: *Physicochemical and Environmental Plant Physiology*, Academic Press, San Diego, 1999.

Oldham P.: *Cost of Production of Major Tree Crops in Dominica*, Roseau, Ministry of Agriculture, 1991.

P. Bhattacharyya and S.S. Purohit: *Organic Farming : Biocontrol and Biopesticide Technology*, Agrobios, Delhi, 2008.

P. Venkateshwara Rao: *Dairy Farm Business Management*, Biotech Books, Delhi, 2008.

Qystein V. Sjaastad: *Physiology of Domestic Animals*, International Book Distributing Co., Delhi, 2005.

R.M. Johri and Sneh Lata: *Taxonomy I : Systematics and Morphology*, Sonali Publication, Delhi, 2005.

S K Singh: *Biotechnology, Plant Propagation and Plant Breeding*, Campus Books, Delhi, 2008.

T N Kataria: *Plant and Crop Physiology*, Pearl Books, Delhi, 2008.

U K Mishra: *Physiology of Animal Growth and Bioenergetics*, Satish Serial Pub House, Delhi, 2008.

Vanderplank, J.E.: *Plant Diseases: Epidemics and Control,* New York, NY, USA, Academic Press, 1963.

White, G.F.: *Natural Hazards: Local, National, Global*, Oxford University Press, New York, 1974.

Yoshida T.: *Cultivation of Citrus Genetic Resources for Evaluation of Characteristics*, Tokyo, JICA, 1996.

Index

❑❑❑